AIRPOWER REBORN

The History of Military Aviation

Paul J. Springer, Editor

This series is designed to explore previously ignored facets of the history of airpower. It includes a wide variety of disciplinary approaches, scholarly perspectives, and argumentative styles. Its fundamental goal is to analyze the past, present, and potential future utility of airpower and to enhance our understanding of the changing roles played by aerial assets in the formulation and execution of national military strategies. It encompasses the incredibly diverse roles played by airpower, which include but are not limited to efforts to achieve air superiority; strategic attack; intelligence, surveillance, and reconnaissance missions; airlift operations; close-air support; and more. Of course, airpower does not exist in a vacuum. There are myriad terrestrial support operations required to make airpower functional, and examinations of these missions are also a goal of this series.

In less than a century, airpower developed from flights measured in minutes to the ability to circumnavigate the globe without landing. Airpower has become the military tool of choice for rapid responses to enemy activity, the primary deterrent to aggression by peer competitors, and a key enabler to military missions on the land and sea. This series provides an opportunity to examine many of the key issues associated with its usage in the past and present, and to influence its development for the future.

AIRPOWER REBORN

THE STRATEGIC CONCEPTS OF JOHN WARDEN AND JOHN BOYD

Edited by John Andreas Olsen

Naval Institute Press
Annapolis, Maryland

Naval Institute Press
291 Wood Road
Annapolis, MD 21402

First Naval Institute Press paperback edition published in 2023.
ISBN: 978-1-55750-103-5 (paperback)

The Library of Congress has cataloged the hardcover edition as follows:
Airpower reborn : the strategic concepts of John Warden and John Boyd / edited by John Andreas Olsen.
pages cm. -- (History of military aviation series)
Includes bibliographical references and index.
ISBN 978-1-61251-804-6 (hardcover : alk. paper) -- ISBN 978-1-61251-806-0 (ebook) 1. Air power. 2. Boyd, John R., 1927-1997--Influence. 3. Warden, John A., 1943---Influence. 4. Air warfare. 5. Strategy. I. Olsen, John Andreas, 1968-, editor of compilation. II. Title: Strategic concepts of John Warden and John Boyd.
UG630.A387 2014
358.4'03--dc23

2014046355

♾ Print editions meet the requirements of ANSI/NISO z39.48-1992 (Permanence of Paper).
Printed in the United States of America.

31 30 29 28 27 26 25 24 23 9 8 7 6 5 4 3 2 1
First printing

Contents

Figures and Tables

Figures

Tables

Preface

Carl von Clausewitz (1780–1831) and other nineteenth-century theorists defined strategy as the use of battle for the purposes of war. Strategy shaped the plan of the war, mapped out the proposed course of different campaigns that composed the war, and regulated the battles that had to be fought in each of the campaigns. In the twentieth century, Basil Liddell Hart (1895–1970) expanded the term beyond its purely military meaning by referring to "grand strategy" rather than the Clausewitzian "military strategy." Liddell Hart suggested that strategy should instead be defined as "the art of distributing and applying military means to fulfil the ends of policy" in which "the role of grand strategy—higher strategy—is to coordinate and direct all the resources of a nation, or band of nations, towards the attainment of the political objective of the war—the goal defined by fundamental policy."

Since then, many have attempted to make the definition even more comprehensive. Alan Stephens asserts that strategy, in its most generic sense, should be defined simply as "the art of winning by purposely matching ends, ways, and means." Such a definition is both classic and modern and serves as a useful starting point for contemporary strategic thought because it encompasses the pursuit of strategic-level *effects* through a logical and coherent matching of the factors that produce those effects. It demands that decision makers identify what they mean by *victory,* and then ensure that their desired *end-state objective* is realistic, clearly specified, and consistent with political objectives. The *ways* chosen to pursue those ends must be feasible, and the available *means* must be suitable and sustainable.

Modern airpower can offer political and military leaders more and better options, provided the underlying strategy links the application of airpower directly to the end-state objectives rather than limiting it to "the battle." Mastering such strategic thought lies at the heart of the military profession, but it requires in-depth knowledge and understanding of theory, strategy, and airpower, and it transcends traditional metrics.

This book, *Airpower Reborn,* builds on earlier theories to offer a conceptual approach to warfare that emphasizes airpower's unique capability to achieve strategic-level *systemic paralysis* rather than tactical-level destruction and attrition of the adversary's ground forces. The five chapters stress the often unappreciated and unrecognized contribution of airpower to conflict resolution, both as an independent force and as an enabler for other instruments of power. Ultimately, the essays in this book acknowledge the essential role of advanced technology in improving airpower capabilities, but they emphasize that air services must cultivate the intellectual acumen of both officers and enlisted personnel and encourage them to think conceptually and strategically about the application of aerospace power. In short, the authors of this volume look beyond the land-centric, battlefield-oriented paradigm that has continued to dominate military theories and strategies long after airpower had offered new options.

I am grateful to my coauthors, all of whom are leading theorists on the subject of airpower as well as outstanding scholars, colleagues, and friends. They have advocated and promoted the significance of airpower and strategy over many years and have helped to educate the new generation of air strategists. I greatly appreciate their willingness to revisit, refine, and extend their previous work to inform a wider audience.

I once more stand in debt to Margaret S. MacDonald for valuable editorial advice; her ability to improve narrative is exceptional. I am grateful to the Swedish National Defence College for sponsoring yet another airpower project. It also has been a true pleasure to work with the Naval Institute Press and its History of Military Aviation series. In particular, I would like to express my thanks to Adam Kane, Adam Nettina, Paul Springer, and Marlena Montagna.

—John Andreas Olsen
Oslo, Norway

AIRPOWER REBORN

INTRODUCTION

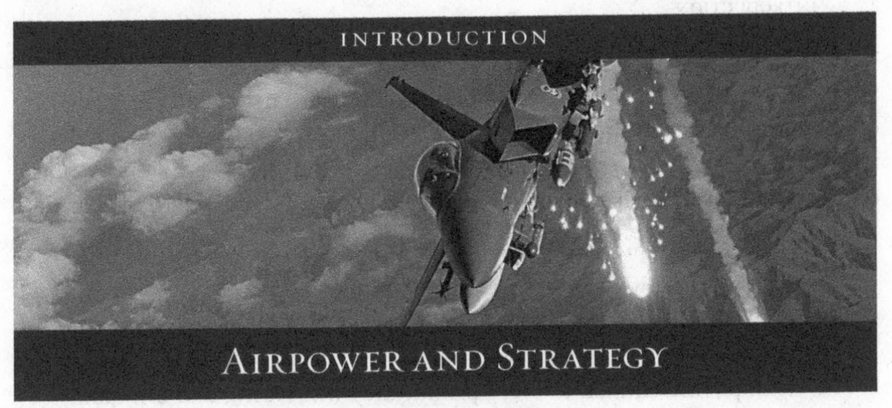

Airpower and Strategy

John Andreas Olsen

From Vision to Reality

Early air theorists who suggested that airpower could be a war-winning instrument confronted an entrenched force-on-force and battlefield-oriented warfighting paradigm. Their enthusiasm about the potentialities of operating above the surface and beyond the trenches led them to make extravagant claims for what airpower could achieve if applied independently, strategically, and decisively against the enemy's vital centers. These visionaries advocated a new way of warfare but could provide no empirical proof of airpower's actual effectiveness in deterring, coercing, or punishing the opponent. Thus the airpower debate became one of faith, interest, and fear rather than an intellectually honest discussion. Tactics and technology, rather than an open-minded appraisal in the broader strategic and operational context, drove the quest to improve airpower effectiveness. As a result, ever since the First World War proponents of air power have had to justify accomplishments against unrealistic expectations rather than against actual results. For most of the past hundred years, skeptics did not distinguish between criticism of airpower and criticism of unsound expectations of airpower.

Airpower has come a long way since the days of Jan Smuts, Giulio Douhet, Hugh Trenchard, Billy Mitchell, John Slessor, and Alexander de Seversky. When airpower finally came of age in Operation Desert Storm (1991), most analysts failed to comprehend why it held such strategic value. The transformation did not result solely from advanced, state-of-the-art technology—such as stealth, stand-off capability, and precision-guided munitions—but from an innovative concept that served as the basis of planning and application. Offensive aerial operations over Iraq showcased airpower as an increasingly powerful and flexible instrument for the pursuit of political objectives, one with continuing relevance well beyond the notion of total war in which it was initially forged.

The campaigns of the 1990s demonstrated that the ability of airpower to deliver the results predicted by pioneers had finally matured. Airpower was the sole military instrument used in Operation Allied Force (1999) and, despite all the challenges encountered in that air campaign, it eventually succeeded in forcing the withdrawal of all Serbian forces from Kosovo. For this reason Sir John Keegan, one of the twentieth century's finest military historians and a lifelong airpower skeptic, concluded that:

> There are certain dates in the history of warfare that mark real turning points. . . . Now there is a new turning point to fix on the calendar: 3 June, 1999, when the capitulation of President Milosevic proved that a war can be won by air power alone. . . . The air forces have won a triumph, are entitled to every plaudit they will receive and can look forward to enjoying a transformed status in the strategic community, one that they have earned by their single-handed efforts. All this can be said without reservation, and should be conceded by the doubters, of whom I was one, with generosity. Already some of the critics of the war are indulging in ungracious revisionism, suggesting that we have not witnessed a strategic revolution and that Milosevic was humbled by the threat to deploy ground troops or by the processes of traditional diplomacy. . . . The revisionists are wrong. This was a victory through air power.[1]

Over the past two decades, airpower has became the "Western way of war"—the preferred military choice for Western political leaders—because it offers the prospect of military victory without large-scale destruction and loss of life. Airpower, however, cannot be decisive or even effective under all circumstances, as demonstrated in a series of counterinsurgency operations at the start of the twenty-first century. The utility of airpower is highly situational. Examinations of the air option and its strategic application in modern warfare must be critical and fair, appraising realistically what it achieved, under which circumstances, and how it compares with other ways of using force.

Fortunately, airpower has become the topic of professional study in its own right, with various institutions offering academic degrees in the field. Richard J. Overy, a professor of history at the University of Exeter, argues that:

> Air power history has become respectable. There is now a body of scholarly academic literature, which has not only succeeded in placing air power in its proper historical context, but in pushing the subject beyond the fighting front to embrace a whole range of different historical issues and approaches. There is now an extensive intellectual history of air power that focuses on the development of doctrine in many differing contexts, thanks in no small part to the interest of air force history offices in understanding the historical roots of current air power thinking. . . . The most distinctive aspect of the new air power history is the growing emphasis on the social, cultural and political dimension of the subject.[2]

Both career members of air forces and other professionals can now build on both research and empirical data to improve their mastery of airpower. Studies by Colin S. Gray, Benjamin S. Lambeth, Richard A. Mason, Phillip S. Meilinger, Richard P. Hallion, Alan Stephens, and many more have contributed to the body of serious, high-quality literature in the field.[3] Although interpretations may vary, the basic facts and the main narrative are now in place.

But as the saying goes at any command and staff college, "It's only a lot of reading if you do it." Moreover, knowledge alone does not a strategist make. While Western defense forces have achieved great success in modernizing their equipment, force structure, and training, that modernization has not extended into strategic thinking. Despite a growing body of scholarly literature on airpower, military doctrine and campaign planning remain ground-centric. Airpower professionals have often proven themselves adept at the technological aspects of war but feel less comfortable in the realm of abstract thought. Thus, current military doctrine governing regular and irregular warfare still centers on war-fighting capabilities rather than on the opponent's overall system and on strategic effects. This stems largely from the still-pervasive belief that only ground forces can ensure military victory and that the enemy leaders will only capitulate when they admit defeat on the "battlefield."[4] Consequently, joint campaign plans, including the aerial component, still favor physical destruction of the adversary's ground forces and elimination of the enemy's military ability to resist. Although airpower has shown itself highly effective at demolishing tanks, artillery, and supplies, this line of thinking imposes severe limitations, since defeating the enemy's armed forces only removes one aspect of the problem. Western strategists must overcome their obsession with "the battle," and instead concentrate on comprehending both the enemy and friendly systems and their leaderships, which represent both the cause of the conflict and the source of any sustainable solution.

The Concept of Systemic Paralysis

To ensure that strategy focuses on ending wars rather than on fighting them, and to avoid reducing strategy to tactics, the authors of this book propose a new conceptual point of departure. In essence, they argue that a viable strategy must transcend the purely military sphere, view the adversary as a multidimensional system, and pursue systemic paralysis and strategic effects rather than military destruction or attrition. Such actions would extend beyond the military battlefield to focus on the opponent's leadership, decision-making processes, and mechanisms for command, control, management, and communication. A leadership-oriented systemic approach identifies and targets centers of gravity, critical vulnerabilities, and key linkages rather than focusing on engaging the enemy through a denial strategy fixated on military forces. Disrupting an opponent's decision-making calculus renders the opponent increasingly deaf, dumb, and blind to proactive and constructive action. How a campaign accomplishes this depends on the situation, but the criteria for success usually include minimal cost in terms of casualties and treasure, minimum damage to the environment and infrastructure, and an end-state acceptable to most parties involved.

Although "systems" are not necessarily mechanical and linear and in fact may be highly complex and adaptive, even an agile and decentralized enemy can still be viewed as a system. An in-depth system-of-systems analysis allows for a broader and all-inclusive approach to affecting key political and physical nodes and connections. Actions that engage centers of gravity, target sets, and individual targets should contribute to achieving predefined desired strategic effects and should set the conditions for follow-on activities such as establishing good governance and state-building measures.

A strategy based on systemic paralysis uses incapacitation to temporarily neutralize the adversary's key elements, break the adversary's cohesion, disrupt the adversary's adaptability, and deprive the adversary of the capacity for timely reorientation. Unable to keep pace with the tempo of events, the adversary's decisions and actions become strategically irrelevant. This concept follows two lines of operations, conducted simultaneously and in parallel: one *process-oriented* to achieve *psychological* impact, and the other *form-oriented* to achieve *physical* impact. The former centers on the intangible—mental and moral—aspects of warfare, while the latter deals with the material sphere. A strategy that combines psychological and physical effects places enormous pressure on the adversary without inflicting unnecessary damage, and it immediately enhances the viability of a follow-on regime.

The ideal of winning without extensive fighting on the ground and at minimal human cost is hardly new—Sun Tzu, Basil Liddell Hart, and J. F. C. Fuller, to mention some, offer extensive advice—but recent improvements in air, space, and cyber power technology open new paths to using resolute military force without deploying large numbers of troops. Here airpower creates significant advantages by using tempo as a strategic quality in its own right. It allows planners to capitalize on the value of time by using the parallel approach in warfare. Only recently has aerospace technology made it possible to attack multiple centers of gravity simultaneously regardless of their locations, to strike them in very compressed time frames, and to control the damage inflicted. Avoiding traditional wars with their perverse and long-lasting impacts reduces postconflict hostility, thus lessening both the suffering and recovery time of the defeated party. Further, the distinction between occupation and control is crucial; a combatant may find it unnecessary and often even undesirable to occupy territory to exercise influence and control. Rather than place large numbers of troops on the ground, in range of enemy weapons, the intervening state can operate from afar. Armed forces need not step into a tactical "red zone" if airpower allows them to dictate strategic and operational effects from a safe distance.

This book is intended to increase understanding of airpower as a political instrument for crisis management and conflict resolution. That understanding will expand and improve if defense professionals can grasp the real strength and uniqueness of the aerial weapon and its ability to produce effects independently from action on the ground and to enable other operations, civil and military. The starting point for a comprehensive strategy should not be technological and tactical deliberations, but a modern and relevant airpower theory: a general set of principles that can serve as a basis for sound and applicable airpower strategies under a given set of circumstances. The ability to link airpower capabilities to strategic thought offers the key to success.

Luckily, strategists need not start from scratch. Two military theorists, both of them U.S. Air Force fighter pilots—colonels John R. Boyd (1927–1997) and John A. Warden (1943–)—have made significant contributions to the evolution of thought about strategic paralysis and effects-based operations. The centerpiece of Boyd's thinking is the OODA (observe, orient, decide, and act) loop; his methodology is process-oriented and his conceptual approaches center on psychological incapacitation of the opponent. His theories emphasize the cognitive and moral spheres of conflict and ultimately involve teaching warriors how to think, that is, teaching the genius of war. By contrast, the foundation of Warden's theories is the Five Rings Model; his methodology is form-oriented and his objective is physical incapacitation of the opponent. Warden's approach is practical and emphasizes the material sphere of conflict, and his conceptual approach centers on neutralizing the tangible resources that enable the enemy to resist. In other words, Warden teaches warriors how to act, that is, to master the principles of war. While Boyd offers a general way of thinking, at times heuristic and esoteric, Warden is prescriptive, identifying both specific target sets and a formula for action. Taken together, their theories create a conceptual framework for imposing systemic paralysis—an alternative to destruction and attrition of ground forces—by lifting strategy beyond operational and tactical levels. In the process, they provide a new vocabulary that challenges the old land-centric terminology.[5] Their approaches serve as an excellent point of departure for thinking of strategy as "the art of winning by purposely matching ends, ways, and means."[6]

SYNOPSIS

Airpower Reborn reflects the advice of Sir Michael Howard to study military affairs in breadth, in depth, and in context.[7] The first chapter presents a historical perspective on airpower theory and airpower strategy, tracing their evolution from the 1920s to the 1980s. The second and third chapters contain in-depth examinations of the strategic concepts Boyd and Warden developed in the 1980s and 1990s, with an emphasis on their contemporary relevance. Chapters four and five provide further context on the theories of these two strategists, with an emphasis on current concepts of operations and enduring principles of airpower. Theory, in this setting, serves as the basic paradigm; strategy represents its generic, mechanism-centered application; and plans of campaign constitute the specific steps for any given situation.

Airpower Theory: Paradigm (to Be) Regained

To illuminate the need for a unified airpower theory, the opening chapter critically assesses how airpower theories evolved during the twentieth century. According to Peter Faber, the development of American airpower theory is a tale of creation, loss, and recovery. Air-minded officers in the 1920s and 1930s attempted to develop a unique theory of air warfare, but during the Cold War the "blue suiters" failed to elaborate and refine the theoretical foundation. In the 1980s and 1990s air strategists finally recaptured some control of their long-lost intellectual destiny, thus ushering in a renaissance in aerospace thought. Faber asserts that this trajectory cannot be understood without recognizing the influence of the dominant, land-centric paradigm of war and the two "languages" used to describe it. The first vocabulary, exemplified by the works of Antoine-Henri Jomini, stems from the Enlightenment and the neoclassical rationalism of eighteenth-century physics and reflects a quest for patterns, predictability, and principles. The second vocabulary comes from the nineteenth-century language of military romantics, exemplified by the writings of Clausewitz, who in *Vom Kriege* stressed the intangibles and uncertainties of war.

From this point of departure, Faber offers an analytic framework to categorize and differentiate among various theories of airpower, from Douhet and Mitchell to Boyd and Warden. His narrative clearly shows that those who sought to promote airpower as an independent and decisive instrument of war had to struggle against a deep-rooted ground-centric tradition, and that their enthusiasm for revolutionary technology influenced to a fault their attempts at theorizing. Faber provides a comprehensive educational template for how to plan, lead, and execute air campaigns through a six-step process: (1) define the desired outcome, (2) identify available capabilities, (3) select a strategy, (4) define target sets and targets, (5) establish cause-effect mechanisms, and (6) address the question of timing. In bringing the debate up to date, Faber characterizes Warden's theories as Jominian in nature, while those of Boyd are far more Clausewitzian. Developing a unified airpower theory for the future—the paradigm (to be) regained—depends on combining these two philosophical strands with insight into airpower strengths and limitations.

Boyd's Way of War: How to Think

With this baseline in place, the second chapter offers a comprehensive presentation, interpretation, and critique of Boyd's work "A Discourse on Winning and Losing," with an emphasis on its current and future relevance to the application of airpower. Professor Frans Osinga examines the formative factors of Boyd's work and suggests that Boyd's approach to strategic thought extends well beyond the well-known OODA loop, a decision-making cycle that in any case is often misunderstood. Osinga asserts that few strategists share the depth of Boyd's focus on the cognitive domain and that Boyd's sophisticated, multilayered, and multidimensional legacy provides strategists with a new set of terms and concepts to study conflict. Admittedly, Boyd's work is at times abstract, strongly biased, and cryptic, but it still offers new avenues into the understanding of warfare. Although Osinga warns against reading too much into Boyd's work as a theory for airpower, he suggests that Boyd's in-depth study of military history and perceptive insights into how social systems learn and behave make him the first postmodern strategist.

Boyd's vision of each party to a conflict as a complex and adaptive system of systems and his characterization of war as a dynamic contest between these systems offer new insight about how to coerce an opponent. Osinga presents a coherent explanation of airpower through the lens of Boyd's "A Discourse," demonstrating how airpower can contribute to military victory when applied in the psychological and morale domains of warfare as well as the physical, and stressing the importance of chance, friction, and other intangibles as inherent components of warfare. In Boyd's terms, language, doctrine, belief systems, experience, culture, symbols, schemata, data flows, knowledge about oneself and the opponent, perception, and organizational ability to learn and change practices—all positioned in the temporal dimension—are at least as valuable as technology, weapons, and the numbers of soldiers in defining combat effectiveness. Boyd's Clausewitzian and process-oriented theories constitute an essential ingredient of the concept of systemic incapacitation through psychological paralysis.

Warden's Way of War: How to Act

In the third chapter John Warden echoes Faber's concern about terminology, emphasizing that airpower can never realize its real potential so long as it is bound to an outdated view of war and its "anachronistic vocabulary." He offers an alternative approach: a new vocabulary and a modern concept of war in which the use of force is associated as directly as possible with end-game strategic objectives rather than with the act of fighting. Warden asserts that leaders must think clearly about four strategic questions: (1) Where do we want to be at some specific point in the future and what are we willing to pay? (2) What are we going to allocate our resources against to create the conditions that enable us to realize the future picture? (3) How, and in what time frame, must we affect the entities against which we will apply our resources? (4) What are our plans for exiting the conflict following success or failure?

Extending his earlier writings on strategy and airpower, Warden advocates a general strategic approach to warfare based on these questions, examines the key elements of such comprehensive strategy, and presents methods that planners can apply in the context of any conflict envisioned. Warden demonstrates how airpower has a unique capability to deliver the effects necessary to realize this transformative vision of system-level warfare. The chapter expands on the concept of the enemy as a system and on the use of the Five Rings Model to identify centers of gravity. Warden uses examples, especially from Operation Desert Storm, to show how the five rings apply in the real world. Warden's Jominian and form-oriented theories aim at systemic incapacitation through physical paralysis; his approach both contrasts with and complements Boyd's line of reasoning. In conclusion, Warden asserts that if airpower is to achieve its full potential, students and practitioners of the air profession must comprehend, believe, and articulate end-game strategy as the foundation of airpower. Failure to do so will constrain airpower, depriving national decision makers of the most cost-effective tool to achieve their objectives.

Fifth-Generation Strategy: A New Modus Operandi

In chapter 4 Dr. Alan Stephens contends that the strategic thinking of Western societies today is largely founded on nineteenth-century dogma of "mass, invasion, occupation, and seizing and holding ground," a first-generation theory that time and again has proven fatally flawed. He suggests that modern-day counterinsurgency theory is "nothing more than a failed variant of first-generation strategy" and strongly criticizes Field Manual 3-24 on Counterinsurgency, referring to it as "the Emperor's New Clothes."[8] Stephens suggests that airpower theorists must put the "broken model" behind them and move toward strategy founded on "knowledge dominance, tempo, precision, and a fleeting footprint." Such a strategy is expressed through operational concepts such as strategic paralysis, strategic raid, rapid halt, no-fly zone, and checkmate by operational maneuver. At the core lies tempo as a strategic quality in its own right. After describing his new approach to warfare, Stephens examines how to apply, adapt, and ultimately master fifth-generation strategy by moving away from faulty land-centric models, based on invasion and occupation by large armies, toward air-centric models that can achieve greater systemic and strategic effects at lower cost in lives, treasure, and damage.

Stephens shares the concerns of the other contributors to this book: that most theorists and strategists use outdated terminology and vocabulary as a basis for conceptualizing warfare, which has proven counterproductive for developing war-winning strategies in a new era. He concludes that the fifth-generation strategy leverages the West's greatest military comparative advantage—airpower—and that recent experience confirms its effectiveness. He speaks for many when he concludes that if Western strategic thinking is to advance, experience indicates that the time has come for a different military culture. Western military leaders must change their conceptual and strategic modus operandi and take advantage of the new capabilities that technology offers.

Airpower Theory: Complete yet Unfinished

In the final chapter, Professor Colin S. Gray offers a general theory of airpower in the form of twenty-seven dicta. As the basis for his conclusions he draws on more than a hundred years' worth of empirical evidence coupled with an understanding of the enduring nature of strategy. Gray suggests that airpower is one of history's most impressive success stories and that the meaning and significance of airpower in warfare are neither mysterious nor controversial; an objective review of a century of airpower history provides an ample basis for a general theory. To Gray, mastery of airpower theory depends on the ability to distinguish between the *nature* and the *character* of airpower, the enduring aspects of airpower as a military-political instrument of national force compared to the ever-shifting technical and tactical elements that express the application of airpower in various contexts of time and place.

Gray also emphasizes the purpose of theory: to offer explanations. In other words, theory serves as a guide on how to approach a situation; it does not provide a checklist formula for successful action. Its true value is measured by its applicability; airpower theory must help operators better serve their political masters. Gray offers insight on the asymmetric advantages of airpower and presents a reasoned assessment of airpower's potential and its limitations. Gray acknowledges Boyd's and Warden's contributions to rebooting strategic thought, praising their focus on the strategic rather than tactical value of airpower and their emphasis on the conceptual rather than the technological dimensions of warfare. Gray's own airpower theory does not represent the final word, but forms the basis for further efforts to better link ends, ways, and means: the political end-state objective, strategy, and airpower capabilities.

Next Steps

Collectively, the authors of this book introduce an approach to "winning" different from the traditional land-centric doctrine that has dominated military thinking from the earliest records of warfare to the present. The concept of systemic paralysis exemplifies strategy as an ideal, but achieving it requires a combination of principles and pragmatism, and so far the strategic value of airpower remains unappreciated and underrated. The essays in this book should foster diverse approaches to strategy by suggesting an avenue of conceptual exploration not constrained by the past, but inspired to formulate prudent strategies for the future.

For most Western nations and their partners, military operations have become more complex, the spectrum of operations has expanded, expectations of airpower have increased, and defense budgets have decreased. To enable their countries to do more with less, the leaders of air forces must become better at linking ends, ways, and means and must fully appreciate the nexus of airpower and strategy. Unfortunately, the curricula of Western air force academies and command and staff colleges offer almost no guidance on strategic airpower thought. To counter this shortcoming, all countries should consider establishing a dynamic and vibrant environment for mastering aerospace history, theory, strategy, and doctrine: a milieu for cultivating broader knowledge of and insight into airpower and a setting in which airpower experts have the opportunity to communicate their narrative to politicians, the media, and fellow officers and to interact to mutual benefit with experts from all sectors of governance. This effort should emphasize the potentially unique contribution of airpower to political objectives and joint operations, and in turn connect to headquarters that do operational planning. Ideas matter.

Such an "airpower project" must have a strategic and conceptual focus, not a tactical and technological one. Countries need to dedicate their "best and brightest" to building an airpower theory for the twenty-first century, capturing it in high-quality, RAND Corporation–like studies, and formulating a serious outreach plan for sharing the findings with politicians, officers, nondefense civil servants, and university faculties. The most effective way to do more with less depends on new ideas and creative thinking; the need for intellectual hubs to craft forward-leaning, air-minded strategic concepts is greater than ever before. Compared to investing in platforms and technology, such an effort involves little expense. All countries can afford to invest in strategic acumen and air-mindedness. This book seeks to help set the direction.

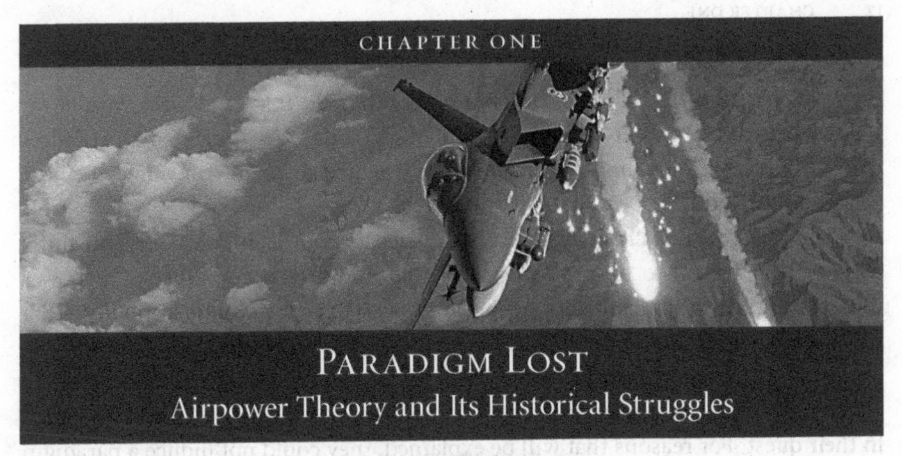

CHAPTER ONE

Paradigm Lost

Airpower Theory and Its Historical Struggles

Peter R. Faber

When Thomas Kuhn published *The Structure of Scientific Revolutions* in 1962, little did he suspect that it would sell, at present count, more than 1.4 million copies. It was a slim volume that soon became a controversial one, particularly because it argued that science was not necessarily "scientific"; it was also a form of ideology. For historians, however, Kuhn mattered for another reason. He introduced the concept of the "paradigm shift," which eventually became a prominent way to map out and explain the past. Periodization—the Age of This, the Age of That—has always been part of the historian's craft, but the virtue of the paradigm shift is that it introduced greater complexity and drama into the presentation of history. At *x* point in time, the argument went, reality and the way societies and institutions interpreted it overlapped; they were in sync. But as time and circumstances evolved, alignment became misalignment. Reality changed, but thanks to sociopolitical dynamics, ideological resistance, institutional imperatives, group dynamics, private self-interest, etc., the way we collectively perceived it did not. For various reasons, societies and institutions attempted to preserve the old paradigm, and the longer they resisted the new one, the more surreal their grasp of reality became. Eventually, the misalignment became so severe that a spasm-like realignment had to occur. The paradigm shift that followed was usually disruptive and disturbing, "epochal" in fact, and led to new realities. New, that is, until another misalignment inevitably occurred and led to yet another shift.

Social scientists and historians soon started to use the paradigm shift as a versatile explanatory and ordering device, but its utility became particularly important to military historians and analysts. It is the foundation, for example, for the vast literature produced on "military transformations" and "revolutions in military affairs" over the past twenty-five years. To cite yet another well-known example, it is also at the center of Philipp Bobbitt's monumental *The Shield of Achilles.* It argues that the nation-state has undergone six paradigm shifts in the modern era, which then yielded six different constitutional orders that fed into, and were created by, six epochal wars.[1] In each instance, the paradigm shift yielded new political orders that subsequently relied upon different bases of legitimacy, until, of course, the next paradigm arrived.

I have begun this chapter with a brief discussion of Kuhnian paradigm shifts because it is central to my thesis: Prior to the mid-1980s, a number of airpower theorists and their sympathizers struggled to overthrow, or even just modify, what had become a dominant, Janus-faced Western paradigm of war. With the possible exception of the United States in the 1950s, when the dominant paradigm and the "Pentomic Army" that tried to preserve it both trembled, these air theorists failed in their quest. For reasons that will be explained, they could not induce a paradigm shift that did not arise naturally, that is, from an at-the-breaking point misalignment between what was "real" and what was perceived. Did air theorists then succeed in "paradigm shifting" after the mid-1980s? I leave that question for the authors of the other chapters in this text to discuss. My focus here is purely historical and it is a tale about "Paradigm Lost."

To support this overall thesis, this chapter has three basic parts. First, I describe the scope and evolution of the dominant, land-centered paradigm of war—Western war—that specific air theorists began to question as early as Giulio Douhet. I cover this in some detail to convey its sheer magnitude and weight, to illustrate how the paradigm became confused with war itself, and to show just how quixotic any attempts to dislodge it proved to be. Since the term "paradigm" is admittedly abstract and dry, however, I often use the term "language" in its stead. In either case, my goal is to describe the brass-knuckled fight for dominance that theorists pursued, both within and outside the dominant paradigm, to promote their preferred visions of war and how it should be fought. Second, I provide a framework that not only helps us differentiate assorted twentieth-century airpower theorists from one another but also provides a template for the development of new theories. Finally, with both a historical context and tool for analysis in hand, I focus on the evolution of airpower theory from World War I up to the period prior to John Warden's *The Air Campaign* (1988). More specifically I discuss thirteen theories developed by air theorists, organizations, and several thinkers who were not air theorists per se but whose writings, intentionally or not, laid out a comprehensible template for the use of airpower.

With this roadmap now in hand, it is important to begin by reminding ourselves that no airpower historian worth his or her salt would ever argue that the creation of theories, strategies, or doctrines occurs in an intellectual vacuum. They all depend on a patrimony that, for better *and* worse, provides them with a paradigm/language they can absorb, reject, transform, or surmount. Airpower theorists did all these things as the twentieth century progressed, but "surmounting" had a special place in their hearts. Unfortunately, what they tried to surmount proved formidable indeed.

The Dominant Language of Modern War

The "scientific" basis of modern Western military theory and strategy had its roots in a text that managed to survive the ravages of time, Publius Flavius Vegetius Renatus' *Epitoma rei militaris* (c. 384–389), popularly known as *De Re Militari*.[2] The reason for its popularity was simple: It was a user-friendly compendium of thirty ancient, largely forgotten military commentators, including Arrian, Frontinus, Polybius, Vitruvius, and others. It was, in short, the late Roman Empire's version of *Warfare for Dummies*.

As a "how to" guide to war, Vegetius' compilation proved irresistible to Renaissance thinkers such as Niccolo Machiavelli, who served as an official in the city-state of Florence from 1498–1512 and used *De Re Militari* as a foundation for his own treatise, *The Art of War* (1521). Machiavelli was not, however, merely interested in restating received wisdom. He sought instead to adapt the old laws of Roman warfare to the new realities of sixteenth-century Italy. He argued this adaptation was possible because human history was immutable rather than unique. The classical military legacy of Rome represented a continuous historical experience that provided infallible and generalizable rules of war that—if applied properly—reduced the relative impact of chance.[3] In other words, because military history was both permanent and knowable, it was an educational tool; it delivered formulaic lessons that inevitably helped one to dissect war rationally.[4]

Based on the recovered wisdom of Vegetius and the updated prescriptions of Machiavelli, the foundation for a scientific approach to war in Europe grew with time.

The Rational, Scientific Language of Modern War

The rigid mechanical-mathematical interpretations of war, which have played such a prominent role in modern military theory, flowered during the Enlightenment. The military rationalists of that era, including Frederick the Great, Henry Lloyd, Heinrich von Bülow, and a collection of lesser lights who went by the collective (and telling) name of Auteurs Dogmatiques, all embraced the linear thinking of the New Physics or Natural Philosophy of the eighteenth century.[5] They categorically rejected the doubtful conclusions of predecessors such as Marshal Maurice de Saxe, who famously characterized war as "a science so obscure and imperfect" that "custom and prejudice, confirmed by ignorance . . . [were] its sole foundation and support."[6] Instead, the soldier-scholars of the Enlightenment embraced the intelligible, mathematical logic of Isaac Newton and his disciples.[7] That Newton was not actually as linear and mechanistic as they thought was beside the point. These men in arms were convinced that reality was "out there"; it was separate and distinct from those individuals who scientifically contemplated the world around them. Consequently, the soldier-*philosophes* argued that it was possible to develop a set of mathematically based maxims to describe and explain a clockwork universe dominated by the Law of Cause and Effect.[8] The law, as it was understood, asserted that the same conditions always produced the same results and that nature was so precise and harmonious that its laws never varied.

Our military "scientists" then went on to claim that within a uniform, cause-and-effect universe, state violence was also knowable and predictable. War was a machine or a mechanism. Vegetius, Machiavelli, and others were right. Warfare was reducible, calculable, and subject to universal and immutable principles. The key, however, remained to identify those "statistical regularities" that actually shaped war.

This task was most famously accomplished by someone who came after the Enlightenment chronologically but who was very much part of its tradition: Antoine-Henri Jomini. Indeed, despite the attempts by some military historians to reverse his popular image as a hidebound Mr. Checklist, at the end of the day Jomini was guilty as charged.[9] He provided a near-endless series of prescriptions on how to approach war. How many factors defined strategy? Thirteen. How many maxims ensured effective lines of operations? Twelve. How many methods existed for effective retreats? Five. And so it went. Through multiple editions of his seminal *The Art of War*, Jomini sought to domesticate organized violence by robbing it of its awful complexity; that is, he attempted to reduce it to its fundamentals. In doing so, his increasingly popular writings reassured skittish European elites that Napoleonic warfare was not a murderous and revolutionary departure from the past. Yes, Jomini admitted, this type of war now involved whole nations in large and disruptive military campaigns, but it was not a blind, mob-driven force of nature that threatened the very foundations of European civilization. Instead, it was part of a rational continuum; it was part of a world of predictable change in which "pure cerebration" still dominated out-of-control will, force, or even luck.

Armed with the proper theory, therefore, those who soberly calculated the ends and means of human conflict would not only succeed, they would also continue to refine war as a science. They would minimize the role of "general friction" and chance, and therefore shape the trajectory of future events, but only if they formalized "patterns from the past in such a way as to make them usable in the present as guides to the future."[10] In other words, "scientific" thinkers such as Jomini agreed with Machiavelli: at its core nothing really changes in history. A "lessons learned" approach to one's military past was both legitimate and helpful. Eternal verities always applied, provided one could express them properly in a rational language of war.

The Irrational or Romantic Language of Modern War

As in other domains, the Enlightenment's insistence that war was largely a science led to a backlash. It had its roots in the broader romantic rebellion of the late eighteenth and early nineteenth centuries, and its seminal military spokesman was Gerhard von Scharnhorst.

The antiscientific approach Scharnhorst used to redefine war had three particularly significant elements. First, the great Prussian reformer repudiated the idea of war as a comprehensible part of a clockwork universe. Instead, it was a blind, demonic force. War was changeable, imponderable, and immeasurable. It roiled with brutal, spiritual energy, and therefore involved a free play of opaque spiritual forces that defied a rigid, one-dimensional, tick-box approach.[11] And since no abstract formula could capture war's sheer diversity, one could not describe it in exclusively mathematical or mechanical terms.

Second, Scharnhorst dismissed the history-has-continuities arguments of the rationalists. He thought Machiavelli and his disciples were wrong: the history of war was not homogeneous and the past did not necessarily repeat itself. Instead, each epoch and the armed violence that characterized it were unique. Indeed, war involved a unique interplay of "possibilities, probabilities, good luck and bad" that militated against historical cycles or patterns.[12] Those who tried to impose personal or absolute frameworks on the past were doomed to fail.

Third, Scharnhorst rejected the idea that nature and therefore warfare were "out there." The world was not separate and distinct from the observer and therefore amenable to objective analysis. Human perception was itself a proactive and creative act; it interacted with the great "out there" to mold and define reality. Human experience, therefore, was a synthesis of the physical and the psychological. The objective world was actually subjective, Scharnhorst concluded, which meant that war was a clash of wills or moral forces unfettered by scientific laws.

If war was demonic, unrepeatable, and a lethal blend of the subjective and objective, did that then mean the rationalist's compulsion to theorize was both erroneous and dangerous? To early romantics such as Gerhard von Scharnhorst and Carl von Clausewitz the answer was "yes," as it was to later ones such as Helmuth von Moltke the Elder, chief of the Prussian General Staff. For all three of these "military romantics" a general theory of war as a "single conceptual system spanning all time" was impossible.[13] Such a theory would inevitably focus on the external forms of armed conflict and not capture the essential "inner nature of circumstances."[14] Further, it would succumb to the siren song of one-size-fits-all maxims and rules. Moltke the Elder, because he dreaded these errors, later argued that those who dared to use hard power needed to depend on *Fingerspitzengefühl* ("fingertip sense") as much as strategy. The latter, by the way, was largely a "free, practical, artistic activity" and a "system of expediencies."[15]

Carl von Clausewitz shared Scharnhorst's and Moltke the Elder's hostility toward compulsive theorizing, but he also muted their absolutist vocabulary. As one of Scharnhorst's disciples, Clausewitz recognized that war was a creative moral act. He rejected strategies of certainty that, as Alan Beyerchen notes, sought "static equilibria, consistent explanations, periodic regularities, and the beauty of symmetry."[16] He further agreed that armed conflict was an intrinsically nonlinear phenomenon. He realized that along with chance, the intangibles, confusions, and dangers of war were part of its essence and not just pesky aberrations one should try to push or calculate away.

As a result, Clausewitz provided multiple and metaphorical definitions of war. War was a continuation of *foreign* policy by other means, by a nation-state that spoke with a single voice. Additionally, war was a game of cards, a duel, an act of commerce, or an act of force designed to impose one's will. Lastly, it was a trinity or interplay of (1) primordial violence, hatred, enmity, and blind natural forces; (2) chance, probability, and the creative spirit; and (3) policy and reason.[17] By providing these diverse definitions of war, Clausewitz illustrated to himself and others that there was an alternative to the rationalist's scientific approach. As a military romantic, he treated armed conflict like a prism. By rotating the prism in his hand and observing the ever-shifting shards of light that flashed before his eyes, Clausewitz could express war's complexity more comprehensively than Jomini and his fellow "scientists" ever did.

Clausewitz did not, however, dismiss the impact of the external, physical dimensions of war. Unlike Scharnhorst and Moltke the Elder, he concluded that these dimensions did introduce broad "statistical regularities" into armed conflict. By examining the phenomenon of war itself and not running after reality-sapping rules of war, Clausewitz decided he could identify its essential elements and yet keep theory grounded in fact. His variety of romanticism kept "theory close to its empirical roots, not letting the language, logic, and polemics of theoretical discourse break away from the untidy, multifarious reality of actual warfare."[18] In short, Clausewitz's model of war lay between the geometry of the Enlightenment and the subjectivity of Scharnhorst and thus avoided many of the self-inflicted wounds generated by both approaches.[19]

To summarize, it is justifiable to say that early nineteenth-century military romanticism served as an antidote to the false universalism and scientific pretensions of earlier and opposing thinkers. Where the axiom writers aimed at fixed values, the romantics postulated that everything in war was uncertain and that strategists had to make their calculations with "variable quantities" in mind; where the scientifically minded emphasized the importance of external and objective forces in defining human conflict, the romantics highlighted the equal importance of psychological forces and effects; and where the rationalists focused on the one-sided, unilateral nature of war against a passive opponent, the romantics posited that war was "a continuous interaction of opposites."[20] By providing a second, competing way of characterizing organized violence, military romantics, whether "pure" ones such as Scharnhorst or pragmatic ones such as Clausewitz, thus restored a balance to previously one-sided speculations about how to use force properly. They weakened, but did not eliminate, the belief that "I-have-the-answer" concepts of war were somehow more truthful or "normal" than those that emphasized the role of intangibles such as irrationality, chance, and probability.

Well and good, some might say, but aside from its intrinsic historical interest, what does the tale of scientific, romantic, or in-between military thinkers really tell us, especially when it comes to the development of twentieth-century airpower theory?

The Dueling Languages of War: Land versus Airpower

As stated at the beginning of this chapter, the above history is important because it illustrates a fundamental point: that it is exceedingly difficult to provoke a Kuhn-like paradigm shift when it does not arise naturally and spontaneously from the conditions around you and when the crushing, multicentury weight of history stands in your way. If by the middle of the nineteenth century Western military theorists had developed fundamental ways of defining and prosecuting wars, they nevertheless represented different sides of the same coin. They reflected, in other words, an open quarrel over a two-part question: "Can states domesticate war to serve their ends and, if so, by how much?" What theorists did not question, naturally enough, was that this quarrel focused uniquely on land warfare, a *type* of war that over the span of four centuries had produced a basic theoretical split (science versus subjectivity) in how military thinkers viewed organized violence. They also took little notice that land warfare had become increasingly confused with war itself. Therefore, when Giulio Douhet and all the airpower theorists who followed in his wake tried to push back against this mighty historical weight—that is to say, when they attempted to provoke a Kuhn-like paradigm shift in military thinking that was largely of their own making—they were stymied.

Being ultimately thwarted, however, does not mean that airpower believers did not put up a good fight. As the twentieth century progressed, a sustained quarrel developed over two updated paradigms of war, one based on the two-part, land-centric vision of warfare described in the previous section, and the other an upstart vision promoted by "aeromaniacs." What the latter were reacting to were assumptions that were incompatible—profoundly incompatible—with their views "about the nature of war and the conditions that produce victory and defeat."[21] The old land-centric vision, for example, assumed that war is a human activity; its essence is defined by intangibles such as will and morale, and that this truth will never change. Second, the vision believed that there are no radical, 90-degree shifts in war; there are only modest changes at the margins. Change, in other words, is evolutionary rather than revolutionary. What may appear to be a sharp paradigm-busting departure from the past is nothing more than normal accelerated change. Third, the old land-centric vision presumed that technology, which lies at the heart of most theories of air warfare, is not a transformative agent in war. If anything, it is just a tool. Fourth, the old vision supposed that territory-centered ground combat will always remain the focal point of war, and tactical or "red zone" operations will remain an inescapable feature of ground combat. Fifth, the traditional vision assumed that airpower cannot substitute for "muddy boots on the ground" because it overpromises. In fact, it has always suffered from an obvious promise-reality gap. Finally, and because of this gap, the old land-centric vision assumed that airpower cannot be a "force of decision" in war.

In contrast to these dominant and unshakeable assumptions about war, what paradigm-shaking view, to cite just three examples, did Giulio Douhet, later Billy

Mitchell, and the members of the interwar U.S. Air Corps Tactical School (ACTS) offer up as an alternative?

The narrative of the "aeromaniacs" often went as follows. For hundreds of years, if you wanted to compel another kingdom or state to change its political-military behavior once and for all, you had to pressure it *indirectly*. Your army had to collide with, knock down, and attrite or defeat an intervening defensive force. Only then would you be free to coerce your opponent in irreversible ways. Needless to say, this time-honored method typically wasted time, treasure, and lives. And to repeat, the approach did not go to the heart of the matter. It did not *directly* change behavior, which meant that time always worked against you. The longer a conflict dragged out, the more variables came into play, which then led to greater amounts of "fog and friction," which then fed war's natural tendency to become total, which then turned the use of hard power, whatever its purposes, into a chaotic melee.

Giulio Douhet, who formalized his thinking with the horrendous cadaver pile of World War I firmly in mind, was the first air strategist who tried to turn this time-honored method of coercion upside down. Instead of relying upon indirect and outside-in methods to compel others, why not use airpower to work from the inside out and strike opponents directly? What we need to do is skip over the "colliding armies" part, he and others argued, and thereby try to neutralize an opponent's willingness and ability to wage war from the very start. This leapfrogging approach then expanded its remit over time. Its proponents claimed that the direct application of strategic airpower could be used to neutralize or even upend the political behavior of an enemy state; it could invalidate its macro-level strategies; it could incapacitate its large-scale political, social, economic, and military systems, as opposed to attacking mere individual targets; it could decapitate a country's civil-military leadership; and it could favor the manipulation of time over the securing of space, thereby robbing an opponent of the opportunity to make cogent, well-informed decisions against you. Indeed, why swat at bees, as the old land-centered thinkers would have you do, when you can shorten the war by destroying the hive? Why merely treat the symptoms when you can radically assault the disease? And further, why ever step into a tactical "red zone" when you can dictate almost immediate strategic and operational-level effects from a distance?

Although traditionalists dismissed them as hopeless ideologues, Douhet and those who followed him actually had specific goals in mind. They wanted, for example, to outdo Clausewitz. They aspired to make airpower a usable political tool instead of a blunt instrument of last resort. They wanted to shorten wars and therefore minimize the cumulative impact of fog and friction. They wanted to make the air weapon, as previously noted, a force of decision. Trapping it into a subservient combined arms framework robbed it of its "natural" potential, they argued.

By making airpower a force of decision, the air theorists further wanted to make war more unilateral and to end the action-reaction cycles that lay at the center of the dominant, land-oriented vision of war and that permitted each cycle to introduce more and more uncertainty into the process. Why not introduce entropy into this collision of wills, our Don Quixotes later asked. Why not use airpower to pummel an opponent at sufficient levels and rates that the action-reaction cycle breaks down, indeed, that it breaks down to the extent that you inflict pain on an incapacitated and disoriented adversary who does not respond but who merely absorbs the punishment you dish out?

From the very beginning, airpower thinkers believed that at least some of these objectives were within their grasp. In their eyes, airpower had unique capabilities. It also operated in a unique, separate, and unobstructed realm. By using an unprecedented technology (at first the strategic bomber) it would free civilian and military leaders from a no-choice reliance on land warfare, which still depended on the moth-eaten means that had lasted for centuries, if not millennia. But to do this, the "aeromaniacs" cautioned, the air weapon had to be used strategically, independently, and directly. And there, as so many analysts have observed, was the rub.

Paradigm Lost: Three Core Reasons Why Airpower's Challenge Went Awry

Thus far, this chapter has looked at the dominant, two-sided paradigm of war that air strategists not only inherited from the past, but that had become so comprehensive and all-consuming that it functioned as a mental straightjacket. The chapter then quickly reviewed some of the mutually antagonistic "truths" that land power and airpower advocates came to believe in over time, colliding truths that amply illustrated that Kuhn's paradigm shifts do not come easily. Finally, the chapter contemplated some of the objectives that certain air theorists still wanted to fulfill. But prior to the later 1980s, they failed. When one then asks why they failed, multiple answers come to mind, but given the paradigm-centered focus of this chapter, three of them should be stressed here.

Problem #1: Military theorists in the modern era have used either the scientific or romantic frameworks already discussed, or they have tried to reconcile the two within a middle ground typically defined by broad, flexible guidelines.[22] This three-part division may seem natural enough, but it raises fundamental questions in relation to airpower theory, not the least of which concerns where to place its practitioners in the spectrum of modern strategic thought. Which language, or combination of the two, did they use to define what was a nascent way of war?

To the detriment of Western air forces, seminal air theorists (Giulio Douhet, Billy Mitchell, Hugh Trenchard, and the "Bomber Mafia" of the U.S. Army's ACTS) and World War II targeting organizations (the Committee of Operations Analysts [COA] and the Economic Objectives Unit [EOU]) did not escape the dominant land-centered vision of war that surrounded them.[23] They used its ideas and they used its words. In short, they developed and promoted didactic, "scientific" strategies. The air theorists and planners had more in common with Jomini and the military-*philosophes* than with Clausewitz and the romantics. Indeed, the air theorists emphasized unilateral offensive action against a largely passive and defenseless enemy; they typically focused on the architectural elegance and calculability of a theory rather than on its accuracy; they inferred that if a theory was symmetrical it must be right, despite the inevitable presence of biases, wishful thinking, and predispositions embedded within its original design; and they deduced theories that despite their scientific pretensions were not necessarily supported by rigorous empirical proof.[24] As a result of these weaknesses, three stubborn pathologies appeared in airpower theory.

Much like their eighteenth-century predecessors, air theorists first sought to develop their own "scientific" model of war that would apply anywhere and anytime. The ACTS Bomber Mafia, for example, adopted "a Jominian, mechanistic view of war—a view of war as a mathematical equation whose variables can be selectively manipulated to achieve success."[25] Therefore, bomber advocates such as Donald Wilson and Frank Andrews argued that any untried theory, including the American theory of high-altitude precision daylight bombardment against the critical nodes of an enemy economy, required "no firmer basis than reasoned logical thinking bolstered by a grasp of the fundamentals of the application of military force and the reactions of human beings."[26] This type of "good deductive reasoning," regardless of how canonical and prescriptive, was acceptable because air theorists defined the world in purely mechanical terms. Maj. Gen. Frank Andrews, for example, noted that modern nations were "as sensitive as a precision instrument." If you damaged a vital part of a watch, the whole ceased to function; so too with a contemporary state.[27] Nino Salvaneschi, an Italian journalist who popularized the ideas of Giulio Douhet, agreed with his American counterparts. He characterized the Great War as no more than a "gigantic watchmaking factory" that was vulnerable to air attack, as did inventor-theorist Count Gianni Caproni, who compared airpower's possible disorganization of Austrian-German war production to breaking a watch by destroying its gears.[28] Again, the clockwork metaphor was a glaring clue that airpower theorists relied on a vision of aerial warfare tainted by the pseudo-scientific approach of the Enlightenment.

But the theorists also suffered from a second pathology: they made a fetish out of quantification and prediction in war. For example, the American authors of Air War Plans Division (AWPD)-1, "the air plan that defeated Hitler," predicted in August 1941 that an initial consignment of 6,860 bombers massed against 125 German target sets would produce victory in six months. In turn, fighter aircraft advocate Claire Chennault predicted in 1942 that he could defeat Japan with 150 fighters and 42 bombers. Last, in early 1964 the Air Force and the Defense Intelligence Agency developed OPLAN 37–64; it anticipated an American victory over North Vietnam in twenty-eight days, provided the United States struck ninety-four "strategic" targets in the north. All three of these "mathematical" examples illustrated a propensity to confuse bookkeeping with analysis, even though analysis is not a "firepower equation writ large" but must include an appreciation of context, combat efficiency, and the intangibles so beloved by let's-keep-airpower-in-its-combined-arms-place traditionalists.[29]

Last, and despite their scientific pretensions, air theorists have always relied on metaphors to buttress the "logic" of their arguments. To reiterate, Count Caproni expressed his opposition to battlefield air operations in a familiar way: "It is not by chasing each bee in a garden that you . . . get the better of the swarm. You should rather destroy the beehive."[30] The ACTS Bomber Mafia, in turn, "proved" the frailty of economic systems by comparing them to either a wispy spider's web or a tottering house of cards.[31] Tear the right strand or pull the right card and the entire structure comes crashing down.[32] This reliance on argument by metaphors, by the way, continued into the 1990s. In his early writings, John Warden suggested that modern societies were closed systems, and therefore vulnerable to collapse, by wrongly comparing them to the human body rather than to the open networks they actually resemble.

Problem #2: If a significant number of the words and ideas used by early airpower theorists came from a long-standing model of war they were supposedly trying to overthrow, and therefore distorted their thinking from within, they certainly had a problem. (The terms they borrowed included the three levels of warfare, interdiction, close air support, fire support, battlefield, and many more.) But by marrying their "infected" model of war to a triumphalist view of technology, they only made things worse. This view of technology as a frictionless "magic bullet" solution for war's problems, incidentally, became particularly prominent in the United States. It flourished, in part, because American Progressivism—a powerful, widespread and technology-friendly social movement cum philosophy—effortlessly reinforced American air strategists' vision of war.[33] Technology was not a mere "force multiplier" in armed conflict, they argued; it was a game changer. Because strategic bombers would, as British prime minister Stanley Baldwin famously said, "always get through," they would also single-handedly usher in a new technology-driven era in war, an era characterized by direct and large-scale effects.

This technological determinism became the U.S. Air Corps' "altar of worship" by the late 1920s, even if it fed into what became a glaring, decades-long promise-reality gap. The "priests" who worshipped at the altar included Edgar Gorrell, William Sherman, Harold Lee George, Kenneth Walker, Ira Eaker, Haywood Hansell, Laurence Kuter, Muir Fairchild, and others. It was these men who in the 1920s and 1930s, either in their writings or while teaching at the ACTS, helped create the concept of high-altitude precision daylight bombardment. They then inculcated this ideology-theory into a number of students who, as senior air commanders, applied their ideas of strategic airpower in World War II.

What did the war teach these men, you might ask? Basically, that they were right. Gen. Carl "Tooey" Spaatz said as much in an article he wrote for *Foreign Affairs* in 1946. Nothing had to change, opined Spaatz. The prewar vision of "leapfrogging airpower" had been vindicated in combat. All the United States had to do now was roll it over into a new nuclear context.[34]

And so it went. The status of strategic airpower waxed in the 1950s and early 1960s, waned during the Vietnam and post-Vietnam eras, and waxed again starting with the publication of John Warden's *The Air Campaign* and the application of John Boyd's trailblazing work within an aerospace power context. The technological determinism of American air thinkers, however, remains to this day. Technology may no longer be tantamount to redemption, but its status still remains inordinately high.

Problem #3: If the "scientific" paradigm of war that air theorists relied upon undermined their program from within, and if their redemptive view of technology made them vulnerable to the time-honored charge that war is an intensely human activity and those who ignore this truth do so at their peril, then their third problem, particularly in the United States, added a thick coat of icing on a proverbial cake. Basically, in advocating their alternative paradigm of war, early American air strategists lashed together a weapon system (the strategic bomber) with a theory (high-altitude precision daylight bombardment against nontactical targets) that required a highly threatening organizational change (the creation of an independent air force). By constructing this unholy trinity, at least in the eyes of land-centric traditionalists, the air theorists' allegedly true motives stood revealed and elicited this response from Gen. John J. Pershing: "An Air Force acting independently can of its own account neither win a war at the present time, nor, so far as we can tell at any time in the future. . . . [If] success is to be expected, the military Air Force must be controlled in the same way, understand the same discipline, and act in accordance with the Army commander under precisely the same conditions as the other combat arms."[35] Make no mistake about what Pershing was saying here: no, you cannot have an independent air force with all the bureaucratic, budgetary, and roles and missions freedoms it implies. American air strategists eventually secured their institutional freedom, but at an exorbitant price. By (1) linking theory, means, and institutional autonomy together, (2) worshipping at the altar of transformative technology, and (3) not totally escaping the long-standing and polluting influence of Napoleonic-industrial warfare, the "paradigm shifters" among them created the impression that they were unreconstructed ideologues bent on a venal mission. That they, in turn, saw their opponents as the living embodiment of "tradition unhampered by progress" goes without saying.

As this section comes to a close, it is important to remember its core theme: air theorists never succeeded in forcing the paradigm shift they wanted. The dominant land-centered paradigm did not get dislodged, even if it was itself divided into scientific and romantic camps. It remained because it was comprehensive and all-consuming and because it was still applicable. It remained because traditionalists developed and deployed a convincing counternarrative to the one developed by air strategists, and because the objectives the air strategists pursued represented, at least initially, "a bridge too far." It remained, as just discussed, because it infected the very alternative air theorists used to try and dislodge it, and it remained because air theorists, at least in the United States, cobbled together a theory, a technology, and an organizational scheme that was tantamount to bureaucratic apostasy. Indeed, prior to the 1980s, artificially induced and permanent paradigm shifts did not come easily, if they ever came at all.

THEORIES OF AIRPOWER: A TOOL FOR ANALYSIS

Having just sketched out the paradigm war at the center of airpower thinking throughout the twentieth century, the next step in this chapter is to offer a conceptual model or tool that can help us identify *past* theories of airpower and shape the development of new aerospace and cyberspace *approaches,* if not outright theories, in the future. The University of Chicago's Robert Pape conceptualized the original model, which Tom Ehrhard and later critics were right to consider too reductive.[36] The model did not, for example, address a full range of aerospace applications, only the severe ones, and it did not consider "the full range of outcomes which strategists seek to achieve or avoid."[37] Despite these valid criticisms, laying out a modified version of the Pape model is still appropriate for three reasons. First, the focus of this chapter is indeed historical, and the overwhelming number of air theories developed prior to the 1980s were particular in their focus and scope. Because the theories advocated using offensive air power in a limited number of ways, using a potentially narrow framework to explain them will not oversimplify or misrepresent them. Second, Pape's model is a "value neutral ordering tool" that emphasizes *process rather than prescription.* Put another way, it remains an empty basket that can accommodate tailored approaches to airpower, as was historically the case, and more expansive ones now. (John Boyd's OODA loop, by the way, is another well-known example of an empty basket, process-centered way to define and prosecute wars.) Finally, Ehrhard improved Pape's model by correcting some of its shortcomings, which is reflected in the following pages.

With the above limitations and justifications firmly in mind, one might do worse than to perform the following six-step process while developing a theory, strategy, or even framework for aerospace power. The process is interactive and works from left to right (from planning to execution) and from right to left (from execution back to planning). As a result, the process is truly practical. It accounts for how things actually work in the real world.

Step One: Desired Outcome

Without exception, a theorist must first ask "what outcome(s) do I want from my use of aerospace forces?" In attacking an opposing state, for example, do I seek political concessions, a military defeat, or an actual change in government? If it is the first option, what particular concessions do I want? Will my opponent make these concessions if put under sufficient duress, or are my political goals unreasonable? If I concentrate on military success, do I want to annihilate or merely neutralize my opponent in battle? Lastly, if I want a change in government, just what type of alternative do I want?

All the above questions are legitimate, but they demonstrate only one type of outcome calculation. As Ehrhard correctly points out, the Doolittle Raid against Japan in 1942 demonstrated a successful application of independent airpower, but its primary goal was to raise domestic morale.[38] The Berlin Airlift was equally successful, but its goal was to check rather than reverse Soviet encirclement. Both examples illustrate that the consideration of outcomes in step one is not a narrow, destruction-oriented wartime activity, nor is it solely preoccupied with the coercion of a hostile state to change its errant ways. The desired outcome could be anything, including economic disruption, changes in domestic or international opinion, continued compliance with civilian policies or doctrines, promotion of confidence-building measures and collective security practices, and creation of legal or moral precedents.

To realize so many different end states, the theorist or aerospace planner should address four essential questions identified by Ehrhard.[39] First, do I seek informal as well as formal outcomes? Second, do I seek short-term rather than long-term outcomes? Third, when I seek an outcome, what interactive impact or changes do I expect domestically, on the "receiver" and on third parties/networks/systems? (The "receiver" can be an international organization, an ad hoc or formal alliance system, a regional block, a nation-state, a nongovernment organization, a terrorist network, a criminal syndicate, etc.) Finally, what factors might the three previous questions bring to bear on my desired outcomes?

Step Two: Capabilities

After establishing your preferred outcomes and updating or replacing them as circumstances change, the next requirements are to (1) gauge the specific politico-military capabilities (i.e., strengths and limitations) of those on the receiving end of your desired outcome(s) and (2) measure the extent of your own ability to project aerospace power. Given that air theorists have historically exaggerated airpower's ability to produce specific outcomes, the determination of mutual capabilities is a vital and necessary step.[40] When determining mutual capabilities, however, the aerospace theorist or planner must define them liberally. After all, many military-oriented factors can help define what a "capability" is, including domestic culture, policy directives, joint military requirements, force structures, readiness and training, available targets, equipment performance, weather, tactics, defensive counters, and general friction.

Step Three: Strategy(ies)

After accomplishing steps one and two, all theorists and planners must then answer a key question before acting. The question, however, can take multiple forms. Pape, for example, asks the following: In my attempts to coerce an opponent, should I pursue a *punishment strategy,* which tries to push a society beyond its economic and psychological breaking point; a *risk strategy,* which tries to do the same thing but gradually rather than all at once; a *denial strategy,* which tries to neutralize an opponent's military ability to wage war; or a *decapitation strategy,* which destroys or isolates an opponent's leadership, national communications, or other politico-economic centers?[41]

Pape concludes in *Bombing to Win* that the strategy best suited for aerial coercion is a specific type of denial strategy, that is, a strategy where you undermine your opponents' confidence in their military solutions, which historically have centered on the control of territory. If you can force your opponents to perform a cost-risk analysis on what it will cost to retain territory, and if you then undermine their confidence in militarily retaining it, then aerial coercion will work. Possession, however, depends on military forces, which means that Pape's coercive strategy for airpower is ultimately a conservative one; you use it to force others to yield territory, and you do that by pummeling their armed forces from the air.[42]

Pape's preferred strategy aside, the four generic ones he suggests for step three are not the only ones to pick from. Air analyst Pat Pentland, for example, asks the theorist or planner to posit a similar and yet different question for this step: Should I adopt a *disabling strategy*, which compromises an enemy's capabilities or resolve; a *delaying strategy*, which uses threats or deterrence to preserve the status quo; or an *enabling strategy*, which tries to create stability where it is weak or does not exist?[43] In terms of using airpower, a disabling strategy includes direct attacks against specific targets. A delaying strategy involves air policing or an air embargo, while an enabling strategy supports military assistance programs.

Pentland understands that as one moves from disruption to stability, military options become less effective, while economic, cultural, and political options become more so. But since Pape's strategies primarily have a high-stakes, wartime focus that involves a recognizable political actor, the virtue of Pentland's last category (his *enabling strategy*) is that it accounts for the growing number of post-1991 non- or quasi-political outcomes that aerospace power tries to accomplish in peacetime. For those who find virtues in both approaches, a mixture of Pape's and Pentland's strategies may offer the best way to secure different end states. But again, the true point here is that picking the right type of strategy to secure your desired ends is an indispensable and required step.

Step Four: Targeting

With preferred outcomes now reconciled to actual capabilities, and with an appropriate strategy or strategies now in place, the theorist or planner should next focus on the critical nexus between the target/objective and the mechanism that will achieve it.

What targets or objectives are the most important? Are they harder to define in peacetime rather than in war? In a high-stakes wartime setting, are the preferred targets enemy leaders; essentials such as oil, information, and electricity; or an opponent's industrial infrastructure, population, or fielded military forces? Are these targets or objectives important individually or in combination? Unfortunately, air strategists traditionally attempted to answer these specific and critical questions before resolving three broader, more fundamental issues.

Issue One: What aspects of an enemy's power should you challenge, either individually or together? As Pentland points out, theorists or aerospace planners could zero in on the *sources* of an opponent's power, which include the military, industrial, or cultural foundations of a state; they could focus on the *manifestations* of an opponent's strength, which include the governmental and ideological projection of force; or they could concentrate on the *linkages* among an enemy's assets, which include the "human and material networks" that determine how effectively a nation organizes and employs its resources.[44]

Issue Two: After theorists or planners review the particular aspects of an enemy's power they want to challenge, they should then consider what generic strategy might work best. They could, for example, adopt a strategy that includes a *direct* approach, which emphasizes head-on assaults against enemy military capabilities; an *indirect* approach, which emphasizes maneuver warfare and the sapping of an enemy's will to fight; and/or a *rapid transition* approach, based on John Boyd's observe-orient-decide-act (OODA) loop, which tries to disrupt or retard an opponent's decision-making calculus in relation to your own, thus making the opponent increasingly deaf, dumb, and blind to your own behavior.

On the other hand, referring back to our previous discussion about competing paradigms, the theorist or planner might adopt an *inside-out* or an *outside-in* approach. In the *inside-out* method, as classically embodied by John Warden's Five Rings Model and every other "strategic" bombing theory of airpower, the attacker strikes vital targets deep within an opponent's territory. Fielded military forces, Warden metaphorically argues, cannot operate effectively without a "brain" directing them. If you sever the brain (enemy leadership) from the body, you incapacitate an opponent from the inside out. The *outside-in* strategy, in contrast, has dominated land warfare for millennia. It necessarily focuses on the forces that surround and protect the inner core of an opposing state. By first eliminating these forces, as we noted before, the planner can then endanger the fountainhead of enemy power.

Issue Three: After determining which generic strategy to adopt, the theorist or air planner might ask, particularly in wartime, "What level of destruction or disruption do I want?" As Kevin Williams observes, a hierarchical series of effects occurs in air targeting. A first-order effect involves the physical or functional destruction of a target within a broader system. If accomplished at a sufficient rate, such destruction yields a second-order effect, which degrades a system's overall ability to operate. An opponent typically responds to this effect by trying to work around it. In a third-order effect, workarounds or substitutions no longer suffice and an enemy nation can no longer compensate for the damage it is experiencing. As a result, the nation must change its military strategy. Finally, a fourth-order effect signals victory, the imposition of your political will on your opponent. You produce this outcome by "achieving third-order effects in a unique and situationally dependent set of target systems."[45] To reach this point, however, air planners must always consider what level of destruction (or disruption) they ultimately desire.

With the above three issues properly resolved—the form or type of power you want to attack, the generic strategy you want to use, and the level of damage you want to inflict—aerospace planners can then move on and determine what specific target set(s) or objectives they want to attack. Carl Kaysen suggests that in doing so they might want to rely on six criteria, particularly when dealing with economic targets.[46]

First, they could consider the military importance of a target. This step might include "a rough classification of the value to enemy military operations of all types of equipment and supplies used by the enemy forces."[47]

Second, the planners might ask, "What proportion of the target is put to *direct* military use?" The higher the proportion, the more important the target may be, especially in short-war scenarios.

Third, they could consider the idea of depth, which measures the military importance of a target in terms of time. "Average depth," according to Kaysen, "is a time concept designed to measure the average interval of time elapsing between the output of a good or service . . . and its appearance . . . in a finished military item in the hands of a tactical unit."[48] Typically, "the measure of depth is important as an indication of the time available to the enemy for the organization of substitute consumption, alternate production, and so forth, before he suffers military damage."[49] Again, in a short-war scenario a target with little "depth" may require immediate attention.

Fourth, the planners should determine the economic vulnerability of a target, which can include the availability of substitutes for processes, equipment, products, or services; the vulnerability of plant layouts and processes; an opponent's resilience; and the ratio of capacity to output.

Fifth, they should consider the physical vulnerabilities of a target set. What type of construction is it? What is it made of? Does it contain additional machinery, stocks of combustible or explosive materials, or other significant items?

Finally, air planners might want to determine the location and size of a target set.[50] Only then is it possible to decide which *specific* targets should be destroyed or disrupted.

Step Five: Mechanisms

After air theorists or planners determine what aspects of an opponent's strength or weakness to assail, what targeting framework to adopt, what order of effects to seek, and what actual target sets or objectives to assault, they must then answer a question that has always had an intimate relationship with targeting: What mechanism(s) do I expect an aerospace assault or operation to trigger? In other words, what changes or outcomes do I expect as a result of a specific action? Will the action, for example, cause economic dislocations, a loss of moral or legal standing, political divisions among allies, a coup, a military retreat, a popular revolt, or just a decrease in the number of political risks an enemy is willing to take? Will it isolate ruling elites from their political base or from fielded military forces and thus cause different types of paralysis?

Unfortunately, our ability to link aerospace ends to desired outcomes remains limited. Over the past ninety years, air planners have become very effective in maximizing "first-order" bombing effects. In fact, decisive physical and functional destruction has become a synonym for targeting efficiency. As stated before, however, the linkage between destruction and outcomes remains unclear. Woven into each theory of airpower are *a priori* assumptions that are neither always obvious nor necessarily wrong. They are, nevertheless, a collection of biases and belief systems more than they are empirical proofs. As a result, air planners have not succeeded, at least historically, in recognizing mechanisms for what they are.

To succeed in the future, such planners first need to define their assumptions closely. Second, they need to keep creating targeting groups that are broadly multidisciplinary in scope and include a variety of civilian specialists. Third, in order to apply the leverage (or mechanism) of aerospace power properly, they should identify "centers of gravity" (COGs) above and beyond traditional target sets. These COGs naturally include political, economic, social, or cultural beliefs and assumptions. They could also include government philosophies, social structures, special interest groups, or demographic factors. Only an expanded appreciation of these types of COGs, and the assumptions behind them, will enable theorists or planners to understand the dynamic relationship between targeting and mechanisms.

Step Six: Timing

As a final step in creating an analytical tool for aerospace power theories and strategies, planners always need to address the question of timing. Given the growing use of aerospace power in peacetime, the marginalizing of large-scale conventional military operations, and the increasing emphasis on asymmetric disruption and paralysis in war, the timing of an aerospace action matters today as it never has before.

Timing, though, has three aspects that need to be considered. First, should a move or assault be sequential, incremental/cumulative, or simultaneous? By answering these questions, the air planner can determine how to use time and space properly. The planner, for example, may choose to conduct a series of measured, escalatory air attacks. If Thomas Schelling is correct, war is a form of vicious diplomacy and has a negotiatory character. The deliberate pauses in a gradualist campaign therefore allow opponents to assess the growing costs and risks of war, which can then lead to proposals, counterproposals, and even course reversals.

On the other hand, and particularly in warfare, the air planner could conduct simultaneous assaults against multiple targets and at different levels of conflict. With the advent and widespread use of advanced data links and precision guided munitions (PGMs) in the 1990s, simultaneous and devastating air attacks became possible. The sheer speed of the attacks could disorganize and confuse an enemy to the point of mental paralysis. As a result, those under attack could capitulate not because of traditional battlefield losses, but because of the disruption of command structures that are unable to cope with the deliberate compression of time and space.

Finally, timing in war can involve a concept little appreciated by past air leaders: a secure reserve force. By withholding a portion of air and space assets from initial operations, and then releasing them while first-echelon forces regroup, aerospace planners can provide a steady stream of pressure in war, rather than traditional waves of pressure.

In conclusion, by focusing on the above six areas—*desired outcomes, capabilities, strategy(ies), targeting, mechanisms,* and *timing*—theorists and planners can avoid the common mistake of fixating on the "how" of air theory, strategy, and even doctrine rather than on the "why." Further, the six areas are not prescriptive; instead they provide the intellectual scaffolding that allows a budding theorist or planner to build new theories and concepts of aerospace employment. To repeat, however, the cause-and-effect relationships among targeting, mechanisms, and outcomes remain under dispute. As in the past, a definitive explanation of this relationship is still the Holy Grail of airpower theory, and the answers offered by multiple theories, whether provisional or not, are very different indeed, as the final section of this chapter will illustrate.

THIRTEEN COMPETING THEORIES OF AIRPOWER

Thus far, this chapter has identified the historical inability of an air-centered paradigm of war to supplant a more dominant and conservative one. Second, the chapter has provided a reality-inclusive, process-oriented model or tool to help distinguish one airpower theory from another and to help develop new theories in the future. The broad aim of this third and final section is to illustrate the use of that model, to compare and contrast ten airpower theories developed prior to 1945 and to explore three airpower theories found "in between the lines" in the works of Irving Janis, Thomas Schelling, and Ernest May. The latter three examples are significant because over three decades—the 1950s through the 1970s—they embodied contemporary debates on how to use airpower effectively. While working through the section, I will describe the thirteen theories and then raise some issues or cautions the reader might want to consider when thinking about them. One of the most important caveats, of course, is that almost all the theories focus on coercing others in conventional conflicts.

Airpower Theory Prior to 1945

In the early days, airpower theorists were divided over one overriding question: What is the best way to persuade an opponent to abandon key political goals and objectives in wartime? At the "strategic" level, the question spawned three schools of thought.

First, there were those who advocated a *punishment strategy,* which sought political concessions or changes in behavior by terrorizing civilians from the air. Members of this group included Giulio Douhet, Billy Mitchell (after the mid-1920s), and Arthur "Bomber" Harris.

Second, there were those who advocated a *risk strategy,* which sought to deprive an enemy of the infrastructure or industrial capacity needed to wage war.[51] Members of this group included Gianni Caproni, Nino Salvaneschi, Hugh Trenchard (in the 1920s), portions of the ACTS faculty, and members of the previously mentioned Committee of Operations Analysts and of the Economic Objectives Unit in World War II.

Third, there were those who advocated a *denial strategy,* which fixated on the destruction of fielded military forces. John Slessor and continental military establishments such as the German and Russian General Staffs supported this option in order to subordinate airpower to what they saw as more pressing ground warfare needs.

When it came to implementing the above options, however, differences of opinion arose. Everyone but Douhet agreed that limited technology restricted air forces to serial attacks. Everyone further agreed that the problems of selecting appropriate target systems, and the specific targets within them, were of overriding importance. As table 1.1 illustrates, however, those who self-selected themselves into one of the above three schools of thought also quarreled over what specific target sets to attack. Only limited agreement existed as to what constituted generic centers of gravity in war and what consequences they would trigger if attacked.

Unfortunately, the men who quarreled so confidently over target sets and mechanisms had limitations in their own theories. For example, those who advocated a terror-centered punishment strategy against an opponent's civilian population either ignored or minimized five major problems. First, rather than spur civilians into political action, aerial terrorism leads to emotional passivity and/or a preoccupation with immediate survival. Second, if civilians believe their government is making a good-faith effort to protect them from air attacks, they will focus their

TABLE 1.1 REPRESENTATIVE THEORIES PRIOR TO 1945

THEORIST(S)	POLITICAL OUTCOME	TARGET SET(S)	MECHANISM
Caproni and Salvaneschi	Military defeat	Major munitions factories	Destroy equilibrium in equipment
Douhet	Change government or its behavior	Population (cities)	Revolution
Mitchell	Change government or its behavior	Vital centers	Civil uprising
Trenchard (1920s)	Change government or its behavior	War materiel, transportation, communications	Operational paralysis
Slessor	Military defeat	Troops, supplies, production	Interrupt or destroy equipment and supplies
Air Corps Tactical School	Change government or its behavior	Key economic nodes (industrial web)	Social collapse, break popular will
Harris	Change government or its behavior	Population (cities)	Fear, lost morale
Wilberg, Weber, German Gen Staff	Military defeat	Enemy field army	Battlefield breakthrough, army destruction
Committee of Operations Analysts	Military defeat	Munitions plants	Material shortages
Economic Objectives Unit	Military defeat	Oil, transportation	Operational paralysis

anger on the attacker rather than on the government. Third, authoritarian regimes are typically indifferent to popular suffering and do not respond easily to domestic political pressure. Fourth, the psychological effects created by air bombardment are often only temporary (because of air defenses and workarounds, populations have time to adapt to their suffering). Finally, the perpetrators of population attacks must systematically ignore existing legal and moral prohibitions against violating noncombatant immunity.

Those who advocated a risk strategy against enemy economies had their own methodological problems up through World War II. For example, attackers tend to confuse the features of their own industrial or economic infrastructure with the critical vulnerabilities of an opponent's system.[52] Second, economies are not brittle. They typically do not have clear breaking points if put under duress. Instead, their vitality ebbs away much like a receding tide, although much less so in an era of PGMs and cruise missiles. Third, air attacks against an economy may provide an indirect way to break a people's will to resist, but the link between economic deprivations, political alienation, and changed political behavior remains unclear. Fourth, attackers continue to have difficulty determining if an economic target is functionally destroyed—if it does not work—as opposed to being physically destroyed. The distinction is an important one when deciding what target sets to revisit.

Last, those who supported a denial strategy against an opponent's fielded military forces collided with a number of biases and assumptions that paradigm-busting American, British, and Italian bomber advocates fervently supported in the pre-1945 years. The latter believed, for example, that aircraft are omnipotent; they can destroy any objective and are invulnerable to any defense. They further believed that command of the air is a necessary and sufficient condition for military victory. They believed that command of the air depends on an air force that is independent of surface forces and is composed of maximum bombing power and sufficient fighting power. They believed an independent air force must already exist, be readily available, always operate in mass and devote all its efforts to autonomous offensive operations. They believed that the proper initial target of an independent air force is the enemy's air force, especially its bases and places of production, and they believed that after achieving command of the air, target selection depends on the context of the moment and requires careful consideration. Finally, they believed that surface forces should just have a defensive function and that an independent, strategically oriented air force should perform the major offensive action.

As if working against the above beliefs did not pose enough of a problem for those who supported land-centered operations, they also had to answer three recurring questions: Who actually controlled air assets in battle? Who decided what roles and missions those assets would perform? And who selected the targets they attacked? To those who believed that army commanders did not understand the value of theater-level air support against targets away from the immediate battlefield, land-centered theories of airpower seemed ill-advised, if not outright heretical.

In addition to having limitations of their own, punishment-, risk-, and denial-oriented air theories also shared certain problems. They were, as already noted, hopelessly "scientific" in outlook. They assumed war was an objective experience in a cause-and-effect universe where one's external means unilaterally impacted another's behavior. As a result of these assumptions, the "how, what, and where" of targeting received much more attention than the "why." No theorist, school, or planning group, despite their pretensions, adequately explained how destroying a particular target set would trigger a specific reaction that yielded a desired political outcome. In other words, if I use *X*, *Y* will happen and cause desired change *Z*. Douhet, for example, ultimately *assumed* that ruthlessly attacking an enemy's population would inspire it to revolt and thus lead a government that cared about the suffering of its people to change or discontinue its political behavior. Salvaneschi and Caproni equally *assumed* that destroying an opponent's munitions plants would create equipment imbalances on the battlefield and thus ensure an opponent's military defeat. John Slessor *assumed* that the dislocation of fielded military forces would yield similar results.

In all cases, however, the pre-1945 theorists-planners ultimately hit an intellectual wall. The targeting process was and remains "civilian" in nature. More than ever, it depends on a variety of academic and professional disciplines. Unfortunately, Douhet and his successors were largely ignorant of politics, economics, cultural anthropology, sociology, comparative religion, and other related fields. Without a holistic, multidisciplinary approach, therefore, their targeting strategies yielded to wasteful trial and error in war, particularly in the sphere of economics. The cause-and-effect relationship between destroying parts of a target system and changing an opponent's politico-military behavior remained fundamentally unclear.

Three Cold War–Era Theories of Airpower

With ten pre-1945 theories of airpower now in hand, which can also be divided into three schools of thought, the rest of the chapter will largely concentrate on three decade-by-decade "theories" indirectly developed by Irving Janis, Thomas Schelling, and Ernest May. In looking at their strengths and weaknesses, it is also possible to see why they were indeed representative of their time.

The 1950s: The Leadership "Theory" of Irving Janis

Air War and Emotional Stress (1951) was a RAND Corporation study conducted by Irving Janis to evaluate the psychological effects of air warfare on civilian populations. By analyzing the emotional responses, attitudes, and behavior of British, German, and Japanese civilians subjected to air bombardment in World War II, Janis drew two noteworthy conclusions about the impact of airpower on noncombatants.

First, the physical magnitude of an air attack was important to those who experienced it directly. Heavy bombing raids, in terms of size and tonnage, temporarily raised political apathy and the distrust of an afflicted population toward its political leaders.[53] Such raids, however, were most effective if they were sporadic and unpredictable, thus depriving an opponent of the opportunity to adapt psychologically for even short periods of time.[54] By disrupting familiar sociopolitical patterns, Janis concluded, irregular air attacks would dishearten an opponent.

Second, Janis concluded that a near-miss experience—a direct exposure to danger or its immediate effects—heightened fear and lowered morale. In Janis' opinion, morale deteriorated most in near-miss groups that narrowly escaped the effects of severe air bombardment.[55] The critical variable here was not the expected level of bombardment, but the degree of one's personal involvement in an air attack. Remote-miss experiences actually calmed fears while intense, terrorizing near-miss experiences appreciably lowered an individual's emotional ability to adapt.[56] In fact, anyone who repeatedly experiences narrow escapes, Janis observed, "may become defeatist, his loyalty to his group may weaken and he may be less willing as a result to work for the achievement of his group's aims."[57] Further, the afflicted person's expectations of victory may diminish, and as his confidence wanes the morale-crushing impact of bombardment would only grow worse.[58]

Janis' discoveries unwittingly provided an alternative approach to Ernest May's factionalism-based strategy, as described below. According to Janis, the heavier an irregular bombardment raid, the higher the number of near misses, which then means the greater the disorganized and maladaptive behavior of those who immediately experience the attack.[59]

Certain factors, however, undermine the utility of heavy and sporadic near-miss bombardment against civilian populations. Negative public attitudes, for example, are not necessarily followed by overt antigovernment behavior.[60] Rather than become contentious advocates of political change, severely bombed civilians typically become docile and depressed.[61] They suffer from the "law of mental inertia," which means they focus on personal survival and cling to the status quo. War weariness may appear, but it is questionable that those who suffer severely will apply political pressure against their government.[62]

If the near-miss option is not particularly effective against civilian populations, would it work when directly applied against an enemy army or leader(s)? In the first case, Stephen T. Hosmer and Group Captain Andrew Lambert, RAF, combined to argue that armies are psychologically coercible through near-miss and direct-hit options if (1) they experience increasingly heavy and frequent bombardment that exceeds their expectations, (2) they are pinned down and isolated, (3) they experience maximum discomfort and fatigue, (4) they develop a sense of expendability and hopelessness by being unable to retaliate, and (5) the enemy offers them a political or military way out of their predicament.[63]

In the case of leaders, J. T. Sink and Tom Ehrhard plausibly argued that Colonel Muammar Gaddafi's near-miss experience with American bombers in 1986 did precipitate a significant—if only temporary—change in Libyan political behavior. Rather than lapse into apathetic inertia, Gaddafi retreated from his overt support of terrorists operating in Europe and the Middle East.[64]

In any case, a Janis-based approach ultimately rests on the beliefs that (1) external threats, to include strategic bombardment, are just as coercive as internal threats; (2) external threats can have a direct impact, rather than an ill-defined indirect impact, on altering enemy behavior; and (3) internal opposition within an enemy nation actually impedes rather than promotes compliance with an external coercer's desires and demands. For those who agree with Janis, near-miss experiences may be a deliberate peacetime policy option, especially given the ready availability of smart munitions and stealth assets.

If the Janis option rests on questionable assumptions, what are they? The weakness of the approach is that it is highly provisional and that the historical record does not support reasonable generalization. The cases available are ambiguous for two reasons: the forces working against compliance may be relatively weak, or additional factors may contribute to a change in behavior.

Thus, on the one hand, one can argue that aerial coercion typically works when external risks are high and internal risks are low or at least not equal to the external ones. Neville Chamberlain, for example, feared the Luftwaffe and yielded to German pressure at Munich. One year later, however, he refused to yield when Germany invaded Poland. His domestic political prestige now depended on preserving Polish sovereignty, regardless of the aerial threat posed by the Luftwaffe.

On the other hand, in some instances aerial coercion succeeds when both the external and internal risks are high. Seen from another perspective, the Libyan raid inspired a coup attempt that, together with the actual attack, may have prompted Colonel Gaddafi to curtail his overt support for international terrorism. Additionally, the February 15, 1991, attack against the Al Firdos bunker, in which U.S. bombs allegedly killed members of Iraqi security forces and their families, may have forced Saddam Hussein to offer a conditional withdrawal from Kuwait and thus avoid the wrath of previously untouched Iraqi elites.[65] Ultimately, the above examples may illustrate that different paradigms lead to different answers.

The 1960s: The Leadership-Population "Theory" of Thomas Schelling

In the 1950s, American strategic bombing theory transformed itself into deterrence theory. As a result, responsibilities were divided between civilian elites and the U.S. Air Force. Strategic Air Command increasingly focused on developing mechanistic targeting plans for nuclear war, while the continued development of strategic theory became the responsibility of civilian strategists such as Bernard Brodie, Herman Kahn, William Kaufman, Albert Wohlstetter, and Thomas Schelling. Schelling, as *Arms and Influence* (1966) confirms, was the Clausewitz of nuclear theorists and the godfather of flexible response, a theory Robert McNamara applied quite unsuccessfully against North Vietnam during the Second Indochina War.

For our purposes, Schelling's theory is best described in chapter 4, "The Idiom of Military Action," of *Arms and Influence.* It remains applicable today, despite the generational revulsion by post–Vietnam-era airmen and women against the gradual use of airpower as an instrument of "vicious diplomacy."[66]

What then is Thomas Schelling's theory of armed conflict, as adapted to the use of conventional airpower? Schelling begins by arguing that airpower is the power to hurt. It is a bargaining chip that is most effective *when held in reserve.*[67] "The threat of violence in reserve," Schelling argues, "is more important than the commitment of force in the field."[68] It shapes the mind and expectations of an opponent, who is reminded that he or she still has something to lose. This process is particularly important in a postnuclear world, where armed conflict has become "a competition in risk taking, a military-diplomatic maneuver with or without military engagement but with the outcome determined more by manipulation of risk than by an actual contest of force."[69] By shaping the cost-risk calculations of an opponent, Schelling hopes to make an enemy behave as he wants. The goal is not to destroy an adversary, but to exact good behavior and prevent further political mischief.

Coercing or compelling an adversary with airpower-based threats, however, depends on several factors. First, any bargaining process requires discrete and qualitative boundaries that both sides can recognize as "conspicuous stopping places, conventions and precedents to indicate what is within bounds and what is out of bounds."[70] Second, all bargaining must be based on actions, or on actions and words, but never on words alone. Third, communications must be simple and form recognizable patterns, except in those limited instances where you want to send a deliberately ambiguous message. If you do not meet these preconditions, Schelling observes, threat-based diplomacy will lack the "high fidelity" it needs to succeed. And if you and your opponent do not communicate in the same "language" or "currency," you both may spin out of control into war.

Wars in a postnuclear world, however, are by definition limited. According to Schelling, the combatants will ultimately commit themselves to some level of mutual restraint. As a result, current conventional wars retain a negotiatory character; they are "a bargaining process, one in which threats and proposals, counterproposals and counterthreats, offers and assurances, concessions and demonstrations, take the forms of actions rather than words, or actions accompanied by words."[71] Yet while the bargaining continues, it is appropriate to deliberately manipulate the tempo of air operations. A gradualist approach, Schelling observes, gives your enemy the opportunity to receive and respond to your signals. Most important, it gives your opponent the opportunity to communicate a willingness to stop fighting, which is the ultimate goal of Schelling's approach.

The failure of gradualism in Vietnam seems to invalidate Schelling's vision of warfare as a vicious form of negotiation and diplomacy. On the other hand, critics of the Rolling Thunder campaign and other measured applications of airpower conveniently ignore a key passage in *Arms and Influence:* "[I]t is so important to know who is in charge on the other side, what he treasures, what he can do for us and how long it will take him, and why we have the hard choice between being clear so that he knows what we want or vague so that he does not seem too submissive when he complies."[72] Since the Johnson administration failed to know these things, it is fair to claim that American policy in Vietnam was a bastardized version of Schelling's vision, rather than the real thing.

Ultimately, the problems with Schelling's theory go further than a misplaced faith in gradualism. If airpower is to succeed as an instrument of vicious diplomacy and signal sending in the future, its proponents must recognize and adapt to at least nine problems. First, large government bureaucracies are not rational, unitary actors. They often lack the necessary subtlety or unity of purpose required to bargain violently for a prolonged period of time. Second, the concept of signal sending wrongly assumes that messages are always clearly given and received. Third, diplomacy based on gradualism allows for adjustments, substitutions, and workarounds by the opponent. Fourth, diplomacy based on gradualism, rather than conveying reasonableness and flexibility, may convey a negative impression; that is, you may appear to lack resolve and/or be politically weak. Fifth, diplomacy based on gradualism does not merely probe the political environment. It alters it and therefore distorts the signals that are sent and received. Sixth, diplomacy based on signal sending wrongly assumes that the actors involved always perform cost-benefit calculations that create identifiable breaking points. Seventh, vicious diplomacy wrongly assumes that governments necessarily care about their people and that they will change their behavior to spare them further suffering. Eighth, vicious diplomacy tends to emphasize tinkering with the status quo. It does not readily involve revolutionary change. Ninth, a long-haul perspective, in any guise, is still not a "natural" characteristic of American diplomacy, vicious or not.

Finally, one has to ask: What is the preferred target set of a gradualist campaign? In Schelling's opinion, it is the enemy population. "Populations may be frightened into bringing pressure on the governments to yield or desist; they may be disorganized in a way that hampers their government; they may be led to bypass, or to revolt against, their own government to make accommodation with the attacker."[73] The result of these assumptions is an air strategy that advocates a gradual assault against an enemy population. The assault may then trigger a cost-benefit calculation by the enemy government that may lead to a change in political behavior. Again, however, the theory assumes that the nine problems listed above will not undermine mutual comprehension and clarity.

The 1970s: The Leadership "Theory" of Ernest May

In chapter V of *Lessons of the Past,* Ernest May focuses on governments that tried to use aerial bombardment to coerce others. He identifies three failures and two successes before 1945. In Ethiopia in the 1930s, Benito Mussolini failed to compel negotiations by aerial chemical warfare. During the Sino-Japanese War, the Japanese conducted three aerial assaults a day to create terror and excite antiwar sentiment in China. Like Mussolini, the Japanese failed to secure a negotiated peace. Finally, fascist air forces in Spain also failed to break republican resistance. In all three instances, the air weapon was an ineffective instrument of political coercion. May does, however, identify two instances in which aerial bombardment did contribute to desired political ends: Italy and Japan in World War II. In both cases bombardment contributed to a change in government and a break with the policies of the past, or so he asserts.

But why did the air weapon have a political effect in the latter two cases and not in the previous three? The answer, according to May, was factionalism. Mussolini routinely pitted various political and bureaucratic factions against each other to retain ultimate power in Italy. At the same time, undersecretaries, bureau chiefs, staff officers, and party functionaries continued to plot against him. They gathered strength while Italy experienced a series of military defeats and a decline in popular resolve. It was at this point that the bombing of Rome occurred. After the attack, two-thirds of the members in the Fascist Grand Council rebuked Mussolini. The bombing also inspired King Victor Emmanuel to dismiss Il Duce, replace him with Pietro Badoglio and a cabinet of nonfascists, and set the stage for a separate peace. According to May, Italy's foreign policy changed because its leadership changed. The actual bombardment of Rome, coupled with the fear of future attacks, contributed to these changes.[74]

The same factors also explain why strategic bombardment succeeded against Japan. A deified emperor ruled over fragmented elites. Civil ministries, political parties, segments of the military, aristocrats, and intellectuals all quarreled with each other. As a result, Emperor Hirohito was the key to change. His decisions to dismiss General Tojo in July 1944 and to surrender in August 1945 were both long overdue. Yet those factions that pressured the emperor to act were like those in Italy; according to May, they were concerned bureaucrats, military officers who distanced themselves from the policies of the past, and politicians working to unseat blameworthy rivals. As in Italy, ongoing bombardment and the fear of future bombardment "had some effect" in changing Japanese foreign policy.[75]

To be fair, May repeatedly claims that his work is more a thought-piece than a theory. He does suggest, however, that nations with fragmented ruling elites are vulnerable to aerial coercion. Strategic air attacks may contribute to the installation of new leaders who have little or no stake in past policies, and thus are willing to change them. But which groups can assume power and adopt new policies? If they exist, can aerial bombardment either strengthen or weaken them? On the latter point, May is largely silent. He suggests that airpower is just one factor among many that coalesce, under historically unique circumstances, to trigger a change in an enemy nation's leadership structure. How airpower contributes to this change depends on the situation and the interpretive skills of the air planner.

May is quite clear, however, about which leadership factions are vulnerable to aerial suasion and capable of dislodging others. Given their access to uncensored information, it is the "pessimists" who best recognize the dangers of a particular policy and thus can agitate for change. They include members of foreign ministries, intelligence bureaus, and internal security forces, as well as ambassadors, cabinet ministers and civil servants concerned with domestic affairs, intelligence analysts, and future forecasters. They also include internal security officers, military leaders not associated with current policies, and politicians eager to secure their own futures.[76]

By analyzing the above factions and how they behaved in Italy and Japan, May draws three conclusions: (1) to reduce an enemy's commitment to a particular policy, multiple levels of a bureaucracy must become pessimistic about its costs, (2) those associated with foreign affairs and intelligence agencies are typically the first to recognize the weaknesses of a particular policy, and (3) the logical advocates for change are those who work in four areas: foreign relations, internal security, intelligence, and even the domestic economy.

Keeping these conclusions in mind, can aerial bombardment promote disaffection between and among bureaucrats and ruling elites? Can airpower nudge dissidents to precipitate a leadership change and therefore a policy change? According to May, the evidence suggests that the threat of bombardment "might contribute to bureaucratic anxiety . . . and hence enhance in some small degree the chances of governmental change resulting in a change in policy."[77] Therefore, as table 1.2 illustrates, a series of air assaults against an enemy nation's leadership might lead to changes in governments or policies. The mechanism of change involves exploiting factionalism within and between politico-bureaucratic elites to create conditions that allow "pessimists" to restructure the government and its policies.

May's endorsement of aerial bombardment as an instrument of coercion acknowledges that change usually occurs, in the words of Albert Hirschman, "as a result of a unique constellation of highly disparate events and is therefore amenable to paradigmatic thinking only in a very special sense."[78] Despite May's keen awareness of the dangers of theorization, however, his provisional model of aerial coercion does raise three issues that should be examined here: (1) the relationship between cause and effect in aerial bombardment, (2) the distinction between external and internal threats to a nation, and (3) the recent impact of modern technology, to include PGMs and stealth systems, on political coercion.

Issue One: How does aerial bombardment, via the unspecified pressures it exerts, yield specific political effects? What *specific* targets do you attack to promote factionalism between bureaucratic pessimists and those who actually dictate policy? How can airpower empower one faction and yet weaken another? On these questions

TABLE 1.2 Three Theories of Aerial Coercion

	POLITICAL OUTCOME	TIMING	TARGET/OBJECTIVE	MECHANISM
Janis	Change policies	Irregular	Leadership or population	Near-miss experience
Schelling	Change policies	Incremental escalation	Population (army; leader)	Future costs and risks calculations
May	Change leaders or policies	Incremental	Leadership	Exploit factions

of cause and effect, and the relationships between the instruments of war and the ends desired, May is silent. Although he tries to link Italian and Japanese shifts in behavior to aerial assaults, the "how" of the process remains unclear. Like his predecessors, May thus *assumes,* rather than empirically proves, that aerial bombardment facilitated the restructuring of enemy leadership elites or reoriented their policies.

May argues, for example, that the restructuring of the Japanese government and its policies began with General Tojo's forced resignation in July 1944. But he also claims that the anti-Tojo conspirators needed more than a year to redirect government policies and sue for peace. Since the timing of American aerial bombardment and the changes in the Japanese government and its policies did not truly coincide, it is impossible to prove or even demonstrate that bombardment had little more than "some effect" in introducing change.

Further, May's cause-and-effect treatment of events depends on two preconditions: that some of the conspirators of 1944 were "moderates" and that they were nonincumbents who disagreed with the direction of Japanese foreign policy. But, as May also admits, the idea of high-ranking "moderates" operating within the Japanese government may be a postwar fiction promulgated by Prince Konoye and other former leaders who wanted to distance themselves from the ruthless colonialism and militarism of the past.[79] As George Quester rightly observes, "May's distinction between incumbents and non-incumbents may stretch political reality a little too much, for many political coups consist of some of the leaders conspiring to oust the rest."[80] Yes, Quester argues, General Tojo fell from power, but "some very powerful elements of the incumbent establishment remained just as powerful afterward."[81]

Ultimately, May's treatment of aerial coercion harks back to Jomini and the rationalists: it is elegant and symmetrical, but its historical foundations are suspect. To his credit, May admits as much. He acknowledges that the cases he analyzes—Italy and Japan—"are few and different and that the evidence is dubious on many points."[82] In fact, it is sufficiently dubious for multiple historians to conclude that strategic bombing did *not* inspire Japanese leaders to reverse course in August 1945. Instead, Russia's entry into the Pacific War, the atom bomb, the U.S. naval blockade, and/or the Soviet invasion of Manchuria dictated the timing of surrender. These interpretations, although a limited sample, confirm that no true consensus exists on why it took more than a year to translate Japan's obvious military defeat into an end to the war. May's claim that newly empowered "moderates" needed more than a year to reverse past policies is only one interpretation among many, yet the relationship he establishes between air attack and political behavior depends on it.[83]

Issue Two: May appears to believe that external threats to a leader's power or survival are less credible and less compelling than threats that come from within his or her own country.[84] In this respect, he seems to agree with Kenneth Waltz, who warns against confusing external means with internal control. According to Waltz, using external force is merely "a means of establishing control *over* a territory, not of exercising control *within* it."[85] Indeed, external force cannot pacify a nation and establish internal political rule. These tasks, Waltz argues, involve processes distinct from those triggered by the external use of force, including airpower. As a result, those theorists who claim that a specific external action (air attack) can lead to a particular internal effect (a change in leadership or policy) are doomed to fail. May tries to avoid this problem, as defined by Waltz, by establishing a hierarchy between internal and external threats.

Indeed, May's hierarchical approach allows him to avoid the "causality trap" and thus sidestep the problem of accurate prediction. The trap equates military power with control and with the ability to coerce others. As Waltz argues, however, equating power with control does not prove that one leads to the other. "To define 'power' as 'cause' confuses process with outcome. To identify power with control is to assert that only power is needed in order to get one's way."[86] Both assumptions, which characterize the thinking of virtually all airpower thinkers, are dangerous. They assert that an intended act and its results are identical. In any scenario, however, the *context* of an action, along with *an opponent's reaction,* will yield unanticipated political results. Because he believes these factors are important, May concludes that faction-driven internal threats are more credible than external threats such as strategic bombardment. The latter can only prod, in some ill-defined way, internal groups to act. As a result, it can only have an indirect effect on what leaders value most: personal survival and the preservation of individual power.

But creating a hierarchy of threats may introduce as many problems as it solves. It remains unclear, for example, just how bombardment indirectly shapes political behavior. Does it actually motivate leadership factions to reverse themselves and reject the status quo, or does it merely provide a pretext for already disaffected groups to act? More importantly, how does bombardment specifically and indirectly increase the influence of one political faction at the expense of another? May cannot answer the last question because of the assumptions he makes. He assumes, for example, that "pessimistic" moderates are more conciliatory than those who cling to the status quo. The hard-liners, in contrast, do not always understand the security problems they face. They misperceive the motives of others, and since offensive strategies such as strategic bombardment are open to a wider range of interpretations than defensive strategies, hard-liners may overestimate the danger posed by external threats. As a result, May argues, they adopt competitive rather than cooperative policies.

Unfortunately, by establishing a simple dichotomy—flexible moderates against inflexible hard-liners—May oversimplifies the relationship between strategic bombardment and political influence. He does not clearly identify or analyze those myriad factors that directly shape an opponent's external policies, nor does he relate them to the indirect impact of strategic bombardment. In other words, he does not ask six basic questions. Who actually controls the foreign-policy-making process? What arguments do competing bureaucratic factions use to support their policy prescriptions? How skillful is each faction in promoting its prescriptions? How well does an opponent understand the attacker's policies and motives? Which interpretation of the attacker's motives dominates the factional debate? And finally, how do domestic factors influence the tug of war over whether to compete or cooperate with an external foe?[87]

Because May does not answer these six questions, his "thought experiment" glosses a key point: the indirect impact of strategic bombardment depends on how enemy elites see their own policies. Strategic air attacks may indirectly tumble hard-liners from power, but only if moderates successfully blame their opponents for the attacks. Therefore, the role of strategic bombardment is twofold: to cause unacceptable suffering and to eliminate any doubt over the root cause of the suffering. If hard-liners successfully portray strategic bombardment as undeserved and as an expression of enemy ill will, their power will actually grow. The air attacks will undercut those who want to change the status quo by making them appear self-interested and disloyal. But if moderate "pessimists" successfully portray the attacks as a reaction to the hostile policies promoted by those in power, it may result in political change. In this case, the hard-liners are not guiltless victims; their wounds are largely self-inflicted. Again, the key is how an opponent sees his or her own policies and how strategic bombardment aids and abets the control of this self-perception by moderates.

Issue Three: Writing in the 1970s, May understandably failed to anticipate the revolution in military affairs that would soon appear. Because of superior command and control technologies, PGMs, and stealth technology, the ability of airpower to cause "bureaucratic anxiety" and either strengthen or weaken the power of "pessimistic" enemy factions may have grown exponentially. As a result, May's causality trap may not be as pronounced as it once was. Precise external attacks may now have a *direct* impact on the internal political dynamics of an enemy state. If so, air planners may exploit factionalism in a different way, as the next section will show.

May's hierarchical theory minimizes an obvious point: a change in leadership and/or political behavior may be the consequence of relative risk calculations made by a leader who sees internal and external threats as *equally* important and credible. In such cases, one can argue that (1) a leader is coercible when the threat against his or her personal or political survival is greater from external sources than from internal sources, (2) a leader is not coercible if the threat from internal sources is higher than that from external sources, and (3) the leader is not coercible when the risks of compliance are equal to the risks of noncompliance.[88] Obviously, these conclusions clash with May's indirect approach. They assume that serious internal challenges to a leader's power and authority will only stiffen his or her resolve.

If the above is true, air planners should *not* indirectly aid and abet disaffected groups of "pessimists." May's "theory" depends on unpredictable and unmanageable enemy factions to promote change. These factions, however, may actually goad an enemy leader to crush them while they are relatively weak. As a result, the air planner should raise the perceived external risks to a leader's personal and political survival to a higher level than the perceived internal risks.[89] The logic of this approach is as follows: if a leader confronts both external and internal threats, he or she will try to reduce those threats over which he or she has the most control, including the elimination of opposition groups. Seen from this perspective, internal factions actually inhibit a leader's willingness to comply with external goals; external demands "*will succeed only when conditions are safe for the leader to be coerced,* and that means lower relative internal risks of compliance."[90] Thus, the external actor must do the opposite of what May recommends. The actor must not rely on unmanageable internal "pessimists" to coerce an opponent to behave properly or allow the perceived internal threats to an opponent rise to a level equal to or higher than that posed by external threats. If they do, calculated intransigence and repression are more likely to occur than a change in leadership or policy.

If internal opposition within an enemy nation actually impedes political change, is there an external, leadership-oriented option available instead? Two immediate options are worth considering.

The first is George Quester's "expectancy hypothesis," as explored by Martin Fracker.[91] The foundations of the hypothesis are (1) the expectations of those who experience an air attack are more important—as psychological variables—than their willing capacity to endure pain, and (2) information that comes as a surprise has greater emotional impact than unsurprising data.[92] With these two propositions firmly established, Quester's expectancy hypothesis suggests that if the ferocity of an air attack exceeds the expectations of those attacked, they might suffer a psychological defeat and abandon future hostilities.[93] Naturally, Quester's hypothesis raises multiple questions, a majority of which also apply to May's provisional theory. What is the threshold that triggers feelings of despair and defeat in an opponent? How much bombardment is too much? Do other factors—cultural values, the nature of the government, political objectives, the dynamics of war termination—raise or lower the threshold of the opponent's resolve? Last, how long will the shock of exceeded expectations last and thus possibly affect political behavior?[94] These questions confirm that the cross-cultural and psychological impacts of air bombardment are still not totally understood. Nevertheless, disrupting the expectations of leadership elites may bear fruit, as might Janis' approach.

A Look Back and a Look Forward to Two Additional Theories of Aerospace Power

As a reminder, this chapter has attempted to accomplish three fundamental tasks. First, it examined the problem of paradigm shifts and the failed attempts by iconoclastic air theorists to provoke such a shift away from land-centric Napoleonic-industrial warfare. There were multiple reasons for this failure, not the least of which was that the new paradigm remained "infected" by the old one. Also, in the last case the air theorists took sides; they embraced the pseudo-scientific characterization of war favored by Jomini and his Enlightenment-era predecessors instead of the military romanticism of Clausewitz and others.

Following this exploration of the paradigm war and the difficulties it caused air thinkers, the second part of this chapter attempted to repair the damage, at least in part. It provided an adaptable, process-oriented model or tool that analysts and planners could use to parse one theory of aerospace power from another and to build others in the future. As noted, they could also use the model/tool to develop theories of employment across the spectrum of conflict, as well as in peacetime operations, coercive diplomacy among states, the modification of international or domestic opinion, or the setting of legal-moral precedents.

Finally, the third part of the chapter demonstrated how to use the framework presented in the previous part. In doing so, it introduced the reader to thirteen different visions of aerospace power, ten of which grew out of the roiling intellectual environment that existed up through 1945, and three of which spread across three decades (1950s–1970s). The theories thus represent a history lesson, a prologue to the more contemporary chapters that follow in this text, a "vocabulary enrichment" exercise for those who need a basic introduction to earlier airpower thinking, and perhaps the partial closing of the previously lamented promise-reality gap.

But does the story really end here? Is it indeed a tale of Paradigm Lost? Is it a tale that features two-fisted Old Testament prophets prior to 1945 and then once-a-decade, "theory"-developing civilians afterwards? Yes and no. After the Second World War, the land-air paradigm war largely ended in the United States. Because it had the unique capacity to do so, the U.S. Department of Defense could resolve this historical problem by accommodating both paradigms.[95] As a result, America's air strategists received their independence in 1947. Beginning in the 1950s, however, their new service started to "fractionate into factions" for at least five reasons. First, the U.S. Air Force did not transform its theory of strategic bombardment into institution-wide doctrine, which left the responsibility to its major commands. Second, Strategic Air Command, under the utilitarian influence of Curtis LeMay, focused more on the "how" of nuclear targeting than the "why" of postnuclear airpower theory. Third, as previously noted, theory became the responsibility of civilian and academic elites rather than "blue-suiters." Fourth, the Vietnam War increasingly split the U.S. Air Force into rival "strategic" and "tactical" camps, and their estrangement only worsened with the development of the Army's AirLand Battle Doctrine in the late 1970s and early 1980s. Finally, and perhaps most importantly, with the introduction of intercontinental ballistic missiles and space-based satellites, the Air Force soon possessed a diversity of ends and means. The previous lashup of theory, technology, and organization—the "unholy trinity" that cast such a long shadow over airpower theory over the decades—no longer applied.

Given all these stresses, the vision bequeathed by those who believed in airpower as a force of decision, and who privileged punishment and risk strategies (people and infrastructures) over those centered on denial (military assets), did indeed start to fragment. By the end of the Vietnam War, in fact, the U.S. Air Force no longer had an integrated, unifying vision. Further, its major commands had become semiautonomous fiefdoms that tied their fortunes to aircraft and weapon systems rather than to something as "irrelevant" as airpower theory. As a result, the Air Force became a collection of mutually suspicious tribes who were suspicious of the institution as a whole and increasingly ignorant of its theoretical underpinnings.[96]

Paradigm truly lost then? Not according to some. Nothing concentrates the mind like defeat, and the U.S. experience in Vietnam was certainly a paradigm-shaking example of that. The misalignment between the reality—it was a hybrid conventional-unconventional war—and the way it was perceived—it was a conflict best mastered by the "American Way of War"—was too great for the U.S. military to ignore.[97] As a result, the search for alternatives to what had been yet another grinding, red-zone-dominated war of attrition soon began, and it inevitably led to John Boyd.

Known as "The Mad Colonel," "Genghis John," or monikers much more colorful than these, John Boyd became a legendary figure to land and air thinkers alike. Basically, anyone from the late 1970s onward who believed in the need to "paradigm shift" the U.S. military (at a minimum), and who agreed with military romantics that all theories and strategies had to account for war's intangibles, cared about Boyd's work. His OODA loop model, originally presented in a sixteen-page paper titled "Destruction and Creation" (1976), looked at conflicts in a thoroughly nonmaterial way. As James Burton described it: "This paper described how the mind goes through the process of analysis and synthesis to form mental concepts that we use to govern our actions as we deal with an ever-changing environment around us. . . . Unless we change our mental concepts as the reality around us changes (destroy the old concepts and create new ones), we make decisions and take actions that are out of step with the real world around us."[98] The quotation may appear to be a plea for "paradigm shifting," which it is, but in time it inspired the idea that decision making is the true essence of war, and if your decisions do not accurately account for the changing environment around you, then bad things will happen. In response, you could focus on yourself and ensure you did not fall into this lethal misalignment, or you could deliberately attempt to manipulate opponents' OODA loops and thereby create mental shocks and concussions in their systems. Since "Decision-Cycle Dominance" became a linchpin of U.S. military thought in the late 1990s, the option the American military chose is self-evident. It opted for a seemingly anti-"scientific," process-oriented theory of war that called for the deliberate manipulation of the fog and friction of war against an opponent.[99] Such an idea, as Frans Osinga explores later in this volume, was not only catnip to the U.S. Marine Corps, it also represented "Paradigm Regained" for iconoclastic air planners who would soon advocate time-compressed, distant air attacks designed to decapitate an opponent's leadership, induce precise system-level collapse, inflict "shock and awe," and much more.

Beginning in the mid- to late-1980s, John Warden became the most prominent thinker among these air theorists. He, though, took a different, more "scientific" approach than Boyd. He never discounted the importance of psychological and political factors in war, but he also argued that the advent of stealth technology and PGMs had made the physical and tangible elements of organized conflict just as important. In fact, one could now put the intangibles—chance and morale, fog and friction—into a distinct category that was separate from the physical. As Warden notes:

> In today's world strategic entities, be they industrial state or guerrilla organization, are heavily dependent on physical means. If the physical side of the equation can be driven close to zero, the best morale in the world is not going to produce a high number on the outcome side of the equation. Looking at this equation, we are struck by the fact that the physical side of the enemy is, in theory, perfectly knowable and predictable. Conversely, the moral side—the human side—is beyond the realm of the predictable in a particular situation because humans are so different one from another. Our war efforts, therefore, should be directed primarily at the physical side.[100]

If Boyd was the first to point toward Paradigm Regained for air strategists—by stressing the central role of paralysis in war, by privileging the manipulation of time over the control of space, and by insisting that these "truths" extended far beyond the battlefield—then Warden and other like-minded air planners (David Deptula, Charles Link, and others) pointed the way by returning to their service's iconoclastic roots. As reflected in table 1.3, Boyd and Warden developed new sets of working propositions that were significantly more optimistic about the controllability of war—through airpower—than the ones that had dominated for decades.

In closing, it is easy to see that these assumptions were controversial and remain so even today. But that is beside the point. By coming from both sides of our Napoleonic-industrial warfare divide—Clausewitz and the romantics versus Jomini and the "scientists"—Boyd and Warden revived what had become a moribund debate. Air-centered utopianism, especially after Desert Storm, forced military establishments to confront yet again an age-old question: What is the most effective way to use armed force? This question needed a context and it was the thinkers and organizations featured in this chapter who provided it. Yes, the more militant among them had their faults. Their iconoclastic, alternative paradigm of war was too "scientific" and even mathematical for its own good, and their redemptive faith in technology was woefully premature. Their fixation on the "inherently offensive" nature of airpower provoked them to underestimate the defensive dimensions of this

TABLE 1.3 Two Paralysis-Centered Theories of Aerospace Power

	POLITICAL OUTCOME	TIMING	TARGET/OBJECTIVE	MECHANISM
Boyd	Yield territory; change policies	Fast tempo	Communications and decision making	Deny strategy/ conflict zone success
Warden	Change leader(s)	Hyperwar: compress time/ space	Leadership + four rings	Decapitation and/or strategic paralysis

domain. Their lashing up a weapon system with strategic-level theories and a demand for organizational independence made them appear bureaucratically "unreliable" and certainly parochial, and their ideological zeal blinded them for decades to the promise-reality gap in their thinking, which enabled others to claim that they lived in an intellectual la-la land of their own making.

But one can also (and easily) argue on behalf of pre-1980s airpower theorists, particularly when it comes to the gadfly, consciousness-raising role they played. First, they reminded their peers that paradigms are not "eternal" and therefore always "real." As Kuhn taught us, reality and its accepted models may coincide for a while, but the models will become outdated or increasingly compromised over time. This truth then impacts how we characterize and respond to war, as it does other human activities. It provokes us, for example, to raise an important question: Should ground combat remain the default option of joint war fighting, if not its outright focal point? Exceptions aside, this bias still exists in America's joint military forces, and it is a questionable one, especially in the long term.

Second, the air theorists largely advocated striking at the strategic-level sources of an opponent's power. In doing so, they reminded us that indirect and lower-level means are inherently wasteful in terms of time, treasure, and lives lost. In contrast, strategic-level aerospace operations against societies and the macro-level resources that sustain them are indeed direct. They can potentially free you from the allegedly "eternal" action-reaction cycles in war (you act, they react), and thereby limit the roles of fog and friction in conventional land-centered wars, which only grow worse as these indirect conflicts drag on.

Third, the air strategists' unbridled faith in technology may have been ahead of its time, but their insistence on its importance reminded others that it was not merely an inert tool. By interacting with those who use it, technology does indeed lead to a complex human-machine dynamic that then changes the nature of war itself. Stressing that war is a human activity, as the opponents of the air theorists repeatedly did, missed or underplayed this point entirely.

Fourth, the "aeromaniacs" unapologetically argued that airpower can be a "force of decision" instead of a mere auxiliary or combined arms tool. Instead of just "preparing" a battlefield, it can shape and sometimes outright define it. To characterize the air campaign in Desert Storm as a preparatory action, for example, grossly misrepresents the environment-defining role aerospace power had in that conflict.

Finally, the more strategic and independent pre-1980s theories of airpower pointed to a slew of possibilities and approaches that are now part of our conceptual landscape. We now think much more seriously about the *effects* high-level targeting creates instead of piling up unusable rubble. We still put a premium on controlling territory, as is often appropriate, but we also privilege the deliberate manipulation of time in ways that permit us to disrupt the decision-making processes and even psychologies of opponents. And because of the ubiquity of stealth and PGMs, military planners can now think of using aerospace power, at least in some cases, as a diplomatic, signal-sending tool instead of just a blunt instrument of last resort.

The pre-1980s airpower thinkers and organizations discussed in this chapter did not clearly articulate all the potential positives I have just listed, but their theorizing provided the intellectual bedrock for the next-step thinking that occurred, as the following chapters will confirm. Indeed, it is because of their restless, paradigm-challenging quest to find alternative ways to use hard power that we should ultimately remember them and should do so, I believe, with respect.

CHAPTER TWO

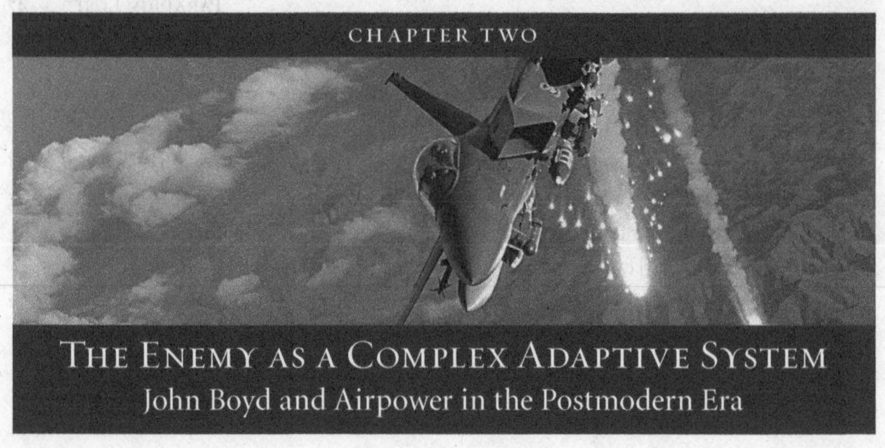

The Enemy as a Complex Adaptive System

John Boyd and Airpower in the Postmodern Era

Frans P. B. Osinga

John R. Boyd has been ranked among the outstanding general theorists of strategy of the twentieth century, along with the likes of Bernard Brodie, Edward Luttwak, Basil Liddell Hart, and John Wylie.[1] As Colin Gray stated: "John Boyd deserves at least an honorable mention for his discovery of the 'OODA [observe-orient-decide-act] loop' . . . allegedly comprising a universal logic of conflict. . . . Boyd's loop can apply to the operational, strategic, and political levels of war. . . . The OODA loop may appear too humble to merit categorization as grand theory, but that is what it is. It has an elegant simplicity, an extensive domain of applicability, and contains a high quality of insight about strategic essentials."[2] Boyd was not an airpower theorist; his work has a much broader scope and aim. Indeed, his work hardly refers to airpower history or theory and mentions airpower only as part of the blitzkrieg concept in support of ground maneuver operations. Scholars must therefore be careful in applying Boyd to airpower theory or employment, because they might overinterpret his work or stretch his ideas beyond what is academically acceptable. With that caveat in mind, this chapter will suggest that the military thought of John Boyd lends itself fruitfully to articulating and explaining the role of airpower in military strategy.

Such a task requires an appreciation of Boyd's work that goes beyond the simple OODA loop construct.[3] This chapter first suggests that despite having some obvious flaws and bearing the hallmarks of his time, his vision is more comprehensive than most people recognize. It contains a consistent line of reasoning as well as many interesting ideas in addition to the OODA loop, such as original formulations of the dynamics of combat and the essence of strategic interactions. Further, his arguments have more substance than most critics recognize. Importantly, the chapter will demonstrate that the popularized concept of the OODA loop is often simply misunderstood. Many regard the tidy graphical depiction of the OODA loop shown in figure 2.1 as *the* concise representation of Boyd's ideas and mistake the simplified model and limited interpretation—rapid OODA looping—for his entire theory. Such

FIGURE 2.1 THE SIMPLE OODA LOOP

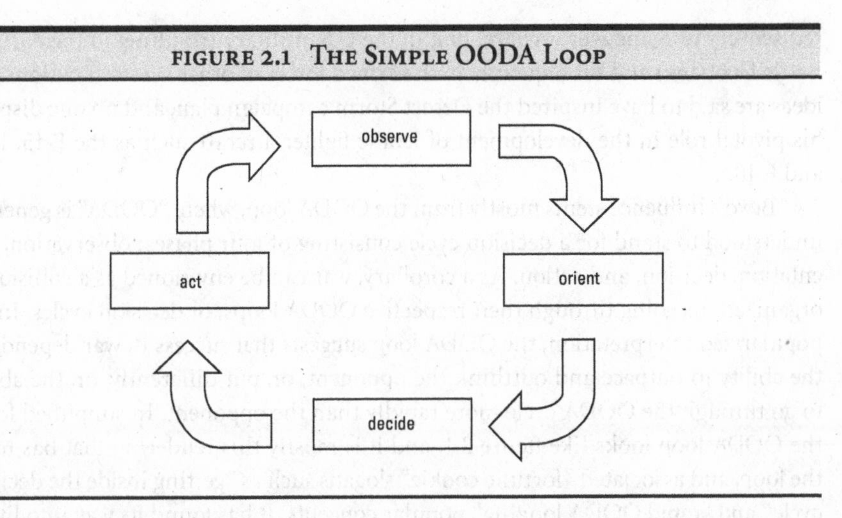

a limited interpretation masks the richness of the concept and the broader theme of Boyd's work: individual and organizational adaptability. That theme is embedded in the metaphor of complex adaptive systems that permeates Boyd's theories, and it is this perspective—considering war, and competition in general, as a confrontation between two (or more) complex adaptive systems—that offers progress in conceptualizing the utility of airpower in war.

This introductory section next presents a brief survey of scholarly views on Boyd, leading into a second part, which explores the key formative factors of Boyd's work. This background is essential to properly appreciate the origins and logic of his arguments as laid out in the often cryptic texts in his various presentations. The third part touches on several key concepts in his work before introducing the "real," comprehensive, OODA loop Boyd drew in 1995, two years before his death. In this model the influence of the nonlinear sciences explicitly comes to the fore. The fourth part highlights key themes in Boyd's work. Drawing upon all the earlier discussion, the fifth part concludes the chapter with a discussion of the relevance of Boyd's work to airpower theory.

INFLUENTIAL, PRAISED, CRITICIZED—AND MISUNDERSTOOD

Boyd's ideas indubitably had a profound influence on contemporary military thought and practice in the Western world. One of his obituaries lauded Boyd as "a towering intellect who made unsurpassed contributions to the American art of war . . . one of the central architects in the reform of military thought which swept the services. . . . From John Boyd we learned about the competitive decision making on the battlefield-compressing time, using time as an ally."[4] Boyd's work fundamentally shaped U.S. Army and U.S. Marine Corps doctrine and contributed to the rediscovery of maneuver warfare, first in the U.S. military (resulting in the AirLand Battle Doctrine) and subsequently in the armed forces of other Western nations. His ideas are said to have inspired the Desert Storm campaign plan, and no one disputes his pivotal role in the development of iconic fighter aircraft such as the F-15, F-16, and F-18.

Boyd's influence stems mostly from the OODA loop, where "OODA" is generally understood to stand for a decision cycle consisting of four phases: observation, orientation, decision, and action. As a corollary, war can be envisioned as a collision of organizations going through their respective OODA loops, or decision cycles. In the popularized interpretation, the OODA loop suggests that success in war depends on the ability to outpace and outthink the opponent, or, put differently, on the ability to go through the OODA cycle more rapidly than the opponent. In simplified form, the OODA loop looks like figure 2.1, and it is mostly this rendering that has made the loop, and associated "fortune cookie" slogans such as "getting inside the decision cycle" and "rapid OODA looping" popular concepts. It has found its way into literature on the concepts of network-centric warfare, cyber warfare, information warfare, and command and control warfare. Indeed, this depiction is often regarded as *the* concise representation of his ideas, and the simplified model that Boyd himself never drew is mistaken for his theory.[5] Over the past two decades the image of the OODA loop has become an icon and has spread like a meme beyond military organizations, infecting business consultants, psychiatrists, pedagogues, and sports instructors. A recent Google search for "OODA loop" produced hundreds of graphic renderings and thousands of hits in articles catering to a wide variety of audiences, academic and otherwise.

The OODA loop has also been the primary target of criticism. The concept is often dismissed as simplistic, overly abstract, lacking the detail that might guide commanders in battle, and too strongly focused on the elements of speed and information superiority, which are of relatively little importance at the strategic level or in, for instance, protracted civil wars or insurgencies.[6] Moreover, the concept might be valid for the domain from which it sprang—air-to-air combat—but inductively suggesting that it also applies to explaining victory and defeat in ground warfare, even at the strategic level, is scientifically and perhaps militarily unsafe. Alternatively, the OODA loop is seen as "argument by analogy" and "pseudoscience," because it is informed by an unwarranted application of theorems from mathematics, thermodynamics, and quantum mechanics to the social world.[7] Some have even asserted that no OODA loops actually exist; in real life command processes look markedly different than the simple iterative model suggests, and many decision-making processes occur simultaneously at various levels in a military organization.[8] Others find that the concept lacks originality, as it seems a mere copy of other models of cognitive and decision-making processes.

Additional critiques note that Boyd's work does not reflect proper academic research, shows a confirmation bias, and never includes hypothesis testing of the OODA loop concept. Admittedly, Boyd never produced a properly referenced scholarly text that presents a coherent argument that can be countered or critiqued. His work consists of a heavily biased overview of military history, based on secondary sources, captured in often cryptic language on slides in a massive presentation and in several shorter presentations that become increasingly abstract. As a theory it is neither original nor comprehensive, as it does not cover nuclear strategy or air or maritime warfare.

A Preliminary Orientation on *A Discourse*

Most critiques (and quite a few of Boyd's critics) never engaged with Boyd's full work, a fate that has befallen the work of other theorists, including Clausewitz. Few are familiar with the source of the OODA loop. *A Discourse on Winning and Losing*, or, as Boyd wrote in the margins of books he studied, "a theory of intellectual evolution," consists of four briefings and an essay.[9] Boyd completed it in 1987, although subsequently he often revised the specific wording on his slides. The essay *Destruction and Creation*, written in 1976, serves as the philosophical foundation for his proposition that uncertainty is a fundamental and immutable characteristic of all aspects of our lives, no matter how good our observations and the theories that derive from them are. He notes that the way to deal with this is to make sure one can recognize the extent to which one's mental model is correct, and the ability to use different models simultaneously. This theme permeated his later work. While working on this essay he also finished a presentation titled *A New Conception of Air to Air Combat*, which did not become part of *A Discourse* but shows a close relation to the essay and foreshadows several ideas he was to explore in *Patterns of Conflict*.[10] It also serves as the conceptual bridge between his fighter pilot background and his maturation as a strategic thinker.

Patterns of Conflict constitutes the historical heart of Boyd's work and is the longest of his presentations. With a first draft completed in 1977 and the final one in 1987, it summarizes Boyd's research on conflict and warfare. It is, in Boyd's own words, "a compendium of ideas and actions for winning and losing in a highly competitive world."[11] It also contains an introduction to the OODA loop. In the presentations *Organic Design for Command and Control* (first draft in 1982) and the one intriguingly titled *The Strategic Game of ? and ?* (first draft in 1986) he uses insights and conclusions from *Patterns of Conflict* to develop arguments about leadership, organizational culture, and the essence of strategy. In Boyd's words, *Organic Design for Command and Control* "surfaces the implicit arrangements that permit cooperation in complex, competitive, fast moving situations," while *The Strategic Game of ? and ?* emphasizes "the mental twists and turns we undertake to surface appropriate schemes or designs for realizing our aims or purposes."[12] The last, very brief, presentation, *Revelation*, "makes visible the metaphorical message that flows from this Discourse."

Later Boyd developed two other briefings fully aligned with, and elaborating, previous arguments. *The Conceptual Spiral*, completed in 1992, presents a different version of the arguments, themes, and insights he had advanced in *Destruction and Creation,* now employed to explain how and why innovation occurs in science, engineering, and technology. The briefing emphasizes that these disciplines can thrive only if they are self-correcting and learning mechanisms. On the other hand, all three disciplines also continuously introduce novelty—and thus uncertainty—into the human environment, and hence Boyd argues that the dynamics at play have universal validity for all types of organizations that strive to survive under conditions of fundamental and unavoidable uncertainty. The final briefing, titled *The Essence of Winning and Losing,* was completed in 1995 and gives a condensed rendering of Boyd's core ideas. Only in this short presentation does Boyd offer a picture of the OODA loop, and then in a much more elaborate rendition than that shown in figure 2.1.

Formative Factors

Strategic theorists, like social science theorists, are affected by intellectual and social factors both internal and external to the discipline.[13] Internal intellectual factors include the influence of schools and traditions of thought on the theorist, including cognitive paradigms, changes in paradigms, and metatheoretical tools. External intellectual factors include ideas borrowed from other disciplines. Internal social factors include the influence of social networks on a theorist's work, while external social factors include the impact of historical change on the structure and institutions of the society being examined.[14] Similarly, the factors shaping strategic theory in a particular period, in a particular country, or by a particular author consist of (1) the nature of war during successive periods, (2) the specific strategic circumstances of the countries involved, (3) the personal and professional experience of the particular thinker, and (4) the intellectual and cultural climate of the period in question.

Not surprisingly, therefore, Boyd's theories are products of his time. Four particular factors had a demonstrable influence on his work: the strategic and defense-political context of the United States in the period in which he developed his ideas, Boyd's professional background, his study of military theory and history, and his keen and evolving interest in scientific developments and the scientific zeitgeist of the time during which he created his vision of military strategy.

Boyd developed his theories over the course of two decades, starting in 1975 and finishing in 1995. His ideas were influenced by, and a response to, the organizational needs of the U.S. armed forces in the traumatic aftermath of the Vietnam War, a period during which the services reoriented themselves toward conventional mechanized warfare tailored for a direct confrontation with the Warsaw Pact in Europe in the strategic context of the Cold War.[15] The Vietnam War had demonstrated stagnation in American strategic and doctrinal thinking and revealed the limits of a military doctrine largely based on the use of nuclear weapons. Additionally, the Arab-Israeli war of 1973 had demonstrated the capabilities of modern Soviet equipment when coupled with an innovative operational doctrine. Finally, Soviet numerical superiority meant that attrition-style warfare was pointless. Boyd's arguments (indeed his bias), including his strong critique of linear and predictable attrition-style battle and his emphasis on the nonphysical dimensions of warfare (unpredictability, initiative, morale, situational awareness, cohesion, agility, speed, maneuver, leadership) must be understood in the context of then-ongoing U.S. Army and U.S. Marine Corps discussions about new doctrine.[16]

Boyd's experience as a fighter pilot in the Korean War, and his subsequent involvement in fighter design, provided the second source of inspiration. Boyd developed a mathematical energy-maneuverability model that allowed him to compare fighter aircraft performances in terms of maneuverability at different altitudes. This model substantiated his impression that American successes in air combat were due to a combination of good training, a cockpit that offered an uncluttered view and thus enabled early detection of enemy aircraft (offering improved "situational awareness"), and the superior ability of American combat aircraft to achieve "fast transients" (rapid change of altitude) compared to the MiG-15, the primary fighter of North Korea. In no small part inspired by Boyd's model and initiatives, the emphasis on maneuverability and adaptability rather than on speed, which had previously dominated fighter design, led to the design of the F-15, F-16, and F-18.

Boyd's extensive study of military history and strategic theories, the third source of influence, familiarized him with different modes of warfare and the constant dynamic of action and reaction, with evolutionary processes. Successful tactics, techniques, and doctrines always harbor weaknesses that an opponent can exploit, which results in specific counterdevelopments and responses that negate the initial advantage. Boyd found confirmation and elaboration of this idea in the works of authors such as T. E. Lawrence, J. F. C. Fuller, Basil Liddell Hart, Heinz Guderian, Sun Tzu, Lenin, Mao Zedong, and Vo Nguyen Giap. Boyd's discussion of military history closely follows their ideas and sometimes their biased views (including those of Liddell Hart and Fuller's interpretation of Clausewitz).

Similar to Fuller and Liddell Hart but less biased, Boyd sought to convince military leaders that the military doctrine and practices of the day were fundamentally flawed. He opposed attrition-type warfare and favored paralysis through maneuver. Also like Fuller and Liddell Hart, Boyd emphasized the moral and cognitive aspects of war. In addition, he echoed Sun Tzu in, for instance, stressing the importance of superior judgment, rapidity, exploiting uncertainty, and constant adaptation as a weapon.

Assertions that Boyd's work lacks clarity have merit, but claims that he merely stated the obvious or restated arguments made by the likes of Liddell Hart miss the mark. True, from the perspective of twenty-first-century military thinking, his work has little originality. Readers can justifiably claim that the first one hundred pages of *Patterns of Conflict* merely "repackage" familiar ideas and arguments. The presentation also reveals some remarkable shortfalls in Boyd's approach to history, which he used (or even cherry-picked) to provide illustrations and empirical validation for patterns he observed in combat, to illustrate his points, and to give his presentations greater credibility.

Boyd's constant reminders about the need to observe the opponent and the wider environment of combat and to act accordingly are not novel ideas, but derive directly from Sun Tzu, as do his ideas about shaping the environment and influencing the opponent before actually engaging in combat. Sun Tzu, too, emphasized using superior judgment and constant adaptation as a weapon. He advocated variety, rapidity, surprise, creating uncertainty and multidimensional warfare, and a host of other concepts and ideas echoed in Boyd's work. Indeed, Boyd deliberately incorporated various concepts from Sun Tzu's work in the first section of *Patterns of Conflict* as a starting point for proposing alternatives to the attritionist style of warfare. At the time, Sun Tzu's work was not mainstream knowledge among military officers, and Boyd made an important contribution by rediscovering, paraphrasing, and updating the concepts of Sun Tzu and the other strategists to suit his own era. To Sun Tzu he adds the blitzkrieg concept, which, according to Boyd, recombines the elements that historically produced success with the new tools of the twentieth century: the tank, the aircraft, and modern communications equipment. To some extent Boyd's rediscovery of operational art was also a novelty. To his audience in the 1970s and 1980s, his insistence on tempo, maneuver, the importance of the moral dimension, organizing in semiautonomous units, etc., was innovative, and his rediscovery of the classical strategists was timely.

Moreover, while his ideas do indeed resemble those of classical theorists of the manoeuvrist school of strategic thought, they transcend them by delving deeper into the essence of victory. Although *Patterns of Conflict* initially suggests that Boyd had merely sought confirmation of his intuitions about the general validity of the rapid OODA loop idea, his study was not so selective or preconceived. While he favored the manoeuvrist style of combat in his discussion of conventional warfare (an understandable preference given his environment), his discussions of the essence of strategy identify the manoeuvrist style of warfare as just one of three possible "categories of conflict." Indeed, he accompanied the discussion of each mode of warfare with a description of the counters to that mode. Thus his thinking reflected the dialectic: the paradoxical and evolutionary character of strategy. Where Liddell Hart saw the indirect approach as a sure path to victory, Boyd saw the processes of action-reaction, of learning, anticipation, invention, and countermovements, and the presentation included an elaborate discussion of cases and causes in which the indirect approach (or blitzkrieg concept) and guerrilla warfare succeeded or failed.

Boyd did not search for one particular optimum solution, but instead acknowledged the contingent nature of war and focused on the universal processes and features that characterize war, strategy, and the game of winning and losing. He led his audience to insights he considered more important: a balanced, broad, and critical view instead of a doctrinaire prescription. This view was inspired by his passionate exploration of an expanding array of scientific disciplines.

The Postmodern Turn: Boyd's Academic Zeitgeist

The scientific zeitgeist is the final formative factor, and it warrants special discussion, although that discussion must necessarily be incomplete.[17] Boyd was the first strategist to introduce the epistemological debates of the 1960s and 1970s into strategic thought and to see the value and consequences of these debates and of other emerging scientific concepts for strategy. He married military history with science, building his theories upon those of Gödel, Heisenberg, Kuhn, Piaget, Polanyi, and Popper, who highlighted the unavoidable uncertainty in any system of thought (as well as the limits of the Newtonian paradigm). From epistemology, cybernetics, and systems theory of the 1960s Boyd ventured into evolution theory, cognitive sciences, and chaos and complexity theory, popularized in a growing number of accessible books published from the late 1970s to the early 1990s. These literatures all converged on the themes of evolution, learning, and adaptation. All offered insights into the functioning of biological and social systems and found their way into political science, sociology, economics, and management theory.

Boyd deliberately introduced concepts from the scientific zeitgeist into the military zeitgeist.[18] In fact, the incorporation of contemporary—postmodern—scientific insights forms an integral part of his contribution to strategic theory, and the multidisciplinary approach to understanding war and warfare became an argument in itself.

The influence of the scientific zeitgeist is already manifest in the 1976 essay *Destruction and Creation* and the presentation *A New Conception of Air to Air Combat*, and it becomes gradually more pervasive and more explicit in the presentations he developed after completing the historically oriented *Patterns of Conflict. The Conceptual Spiral*, completed in 1992, refers exclusively to sources and illustrations from science and technology and repeats the argument of the essay. In his last presentation, *The Essence of Winning and Losing* (1995), Boyd unfolds in just a few slides a highly conceptual synthesis of his work and presents his OODA loop in abstract terms where the central themes of the scientific literatures return.

Boyd's intellectual education between 1960 and 1995 was an important period for science, philosophy, and culture, because in this period a "paradigm shift" occurred in the natural sciences and by extension in the social sciences and culture as well. The shift constituted a movement away from a scientific worldview based on Newtonian, linear, analytical, objectivistic, reductionist, deterministic, and mechanical concepts toward a new mode of scientific thinking based on concepts such as entropy, evolution, organism, indeterminacy, probability, relativity, complementarity, interpretation, chaos, nonlinearity, change, complexity, and self-organization. Instead of nineteenth-century physics, quantum mechanics and biology are key disciplines.

This chapter describes all of which appear on Boyd's reading list and in his work: epistemology and cognitive sciences, evolutionary biology and ecology, quantum mechanics and relativity theory, cybernetics and information theory, and chaos and complexity theory.

These concepts feature prominently in Boyd's work. Boyd's particular concern with and take on epistemology, the centrality in Boyd's work of the factor of uncertainty as well as the need for a multitude of ideas and perspectives for understanding and shaping action (multispectrality), and his deconstructive method of pulling perspectives apart and looking for new possible connections and meaning, are shared with postmodern theories such as constructivism, deconstructionism, and structuration theory. Indeed, knowledge production is at the heart of both Boyd's work and in postmodernism.

Neo-Darwinism, Evolutionary Epistemology, and Military Strategy

The concepts discussed above feature explicitly as points of departure for Boyd's work. Indeed, Boyd's opus starts with an epistemological investigation permeated by these notions, an effort to understand the process of "intellectual evolution," as he scribbled in various books he read. In *Destruction and Creation* he elaborates on the insights of Kuhn, Popper, and Polanyi, who all described the generation of knowledge concerning an uncertain and ever-changing world and convinced him that uncertainty is as inescapable as thermodynamics, quantum mechanics, and the uncertainty principle. As Boyd notes in several presentations, "we cannot determine the character and nature of a system within itself and efforts to do so will only generate confusion and disorder." According to Boyd, individuals and organizations must deal with the key challenge posed by this fundamental uncertainty if they are to make meaningful decisions and take appropriate actions that ensure their survival and growth. As a corollary, Boyd proposes leveraging uncertainty as a tool in competition and warfare.

In Popper, Boyd read that science is an evolutionary process in which theories compete to offer the best explanation of phenomena through the process of hypothesis and test. Polanyi, along with Heisenberg, demonstrated that discoveries result from highly subjective processes of observation. The design of an experiment determines what slice of reality will be examined, and the act of observation shapes reality. Moreover, as Polanyi added, reality is too complex and vast in scope for any one researcher to grasp in full, so scientific progress, and improved understanding of our environment, come from scientists sharing diverse insights. A full grasp of reality requires multiple perspectives, which in turn demands an open scientific community if science is to prosper. All the authors Boyd studied in this regard agreed that knowledge is always incomplete and uncertainty unavoidable, and that improved understanding demands constant development and improvement of theories and schemata (or, in Boyd's terms, mental models and mental constructs).

The OODA loop concept incorporates these insights. In a sense, it seems like a graphic rendering of Popper's "normal model" of scientific research (the hypothesis-test process). Popper sees strong similarities among the processes of adaptation at play in genes, science, and human behavior; he regards culture and genetic heritage as prime determining factors driving adaptation, factors also included in Boyd's drawing of the OODA loop.[19] In *The Conceptual Spiral* Boyd offers his definition of science as "a self-correcting process of observation, hypothesis, and test," and the comprehensive OODA loop explicitly includes the notion of hypothesis testing.[20] The more comprehensive drawing of the loop, however, suggests that Polanyi's work, and particularly Kuhn's response to Popper's theories with regard to how science advances, how scientists learn, and how knowledge grows, had the greatest influence on Boyd's ideas. Kuhn taught Boyd that progress—in the form of improved understanding of real-world phenomena—depends on the Popperian notion of problem solving within a paradigm as well as the Kuhnian emphasis on discovering mismatches between a paradigm and reality.

These notions already appear in *Destruction and Creation.* Directly referring to Kuhn's 1970 edition of *The Structure of Scientific Revolutions,* Boyd includes a rendering of the Popperian dynamics of science within a paradigm: "[T]he effort is turned inward towards fine tuning the ideas and interactions in order to improve generality and produce a more precise match of the conceptual pattern with reality. Toward this end, the concept—and its internal workings—is tested and compared against observed phenomena over and over again in many different and subtle ways." But Boyd echoes Kuhn by stating that at some point "anomalies, ambiguities, uncertainties, or apparent inconsistencies may emerge."[21] In subsequent presentations Boyd often refers to the concept that a paradigm—a closed system of logic—cannot be disproven from within; therefore, one must look across various systems to ascertain the nature of reality and evolve valid orientation patterns. The necessity of applying multiple perspectives (in command centers as well as in scientific inquiry) comes to the fore when Boyd asserts in *Organic Design:* "Orientation is an interactive process of many-sided implicit cross-referencing projections, empathies, correlations and rejections . . . expose individuals, with different skills and abilities, against a variety of situations—whereby each individual can observe and orient himself simultaneously to the others and to the variety of changing situations."[22]

Strategic activities thus resemble the Kuhnian scientific endeavor. Indeed, Boyd asserts, *The Conceptual Spiral* also represents "A Paradigm for Survival and Growth." Boyd reinforces the neo-Darwinist slant in this sentence at the beginning of *Destruction and Creation,* stating that "studies of human behavior reveal that the actions we undertake as individuals are closely related to survival." Individuals seek "to improve our capacity for independent action." In the opening slides of *Patterns of Conflict* this theme resurfaces when Boyd discusses human nature: again, the goal is "to survive, and to survive on one's own terms, or improve one's capacity for independent action." Due to forced "competition for limited resources to satisfy these desires, one is probably compelled to diminish adversary's capacity for independent action, or deny him the opportunity to survive on his own terms, or make it impossible to survive at all." For Boyd, "Life is conflict, survival and conquest" (a statement that reveals the shadow of the 1973 oil crisis and the gloom in the influential *Report of the Club of Rome,* which warned of the depletion of the earth's resources). Therefore, instead of relying only on insights from military history, in studying war "one is naturally led to the Theory of Evolution by Natural Selection and The Conduct of War." Darwin also features in *Organic Design for Command and Control,* where Boyd states that "we observe from Darwin that the environment selects [and] the ability or inability to interact and adapt to exigencies of environment select in or out."[23]

Cognitive Sciences

In parallel with the literature on epistemology, Boyd perused a steady stream of new studies on the functioning of the brain. Survival and growth hinge upon adaptive cognitive processes, which would necessarily have a bearing on strategic thought and the processes involved in commanding military units. The cognitive revolution, combined with neo-Darwinist studies and later with complexity theory, showed Boyd the role of schemata formed by genetics, culture, and experience.[24] Indeed, his model can in part be traced to Gregory Bateson, who developed a model of mind based upon system-theoretical principles.[25] Bateson asserted that "if you want to understand mental process, look at biological evolution and conversely if you want to understand biological evolution, go look at mental process."[26] The work of Jacques Monod, in turn, told Boyd that we observe and respond through a genetically determined program, which results from "experience accumulated by the entire ancestry of the species over the course of its evolution," which in turn shapes action and current experience after analysis of the effect of the action.[27] The human nervous system interacts with the environment by continually modulating its structure, and emotions and experience play a large role in human intelligence, human memory, and decisions. The influence of culture was highlighted by psychologist Jean Piaget, neo-Darwinists Monod, Dawkins, and Wilson, and Edward Hall, whose book *Beyond Culture* taught Boyd that "everything man does is modified by learning and is therefore malleable."[28] But once learned, "these behavior patterns, these habitual responses, these ways of interacting gradually sink below the surface of the mind and, like the admiral of a submerged submarine fleet, control from the depths."

The poststructuralist influences are also evident in the description of the orientation element of the OODA loop, when Boyd asserts that "Orientation, seen as a result, represents images, views, or impressions of the world shaped by genetic heritage, cultural traditions, previous experiences, and unfolding circumstances."[29] The following examples, from *Organic Design for Command and Control,* also reflect this concept: "Orientation is the Schwerpunkt. It shapes the way we interact with the environment—hence orientation shapes the way we observe, the way we decide, the way we act. Orientation shapes the character of present observation-orientation-decision-action loops—while these present loops shape the character of future orientation."

These insights led Boyd to conclude that "We need to create mental images, views, or impressions, hence patterns that match with activity of world." They gain a strategic character when Boyd advances the idea that "We need to deny adversary the possibility of uncovering or discerning patterns that match our activity, or other aspects of reality in the world."[30] In *The Strategic Game of ? and ?* Boyd discusses ways to maintain interaction with the ever-changing world to ensure one's ability to adapt. He asserts that humans accomplish this "by an instinctive see-saw of analysis and synthesis across a variety of domains, or across competing/independent channels of information, in order to spontaneously generate new mental images or impressions that match up with an unfolding world of uncertainty and change."[31]

This view also reflects the ideas of Conant, whose writings taught Boyd the benefits and dangers of the theoretical-deductive and empirical-inductive approaches, and that "without a combination science does not progress."[32] The reconciliation of both approaches is essential to the continuation of a free society in an age of science and technology.[33] Subsequently, Boyd would recommend that military strategists apply both analysis and synthesis, and both induction and deduction, to comprehend the world and an opponent's system. This mode of thinking, in line with the ideas of Polanyi, became a key insight that Boyd wanted to convey as an essential element of proper strategic thinking.

Open Systems

Boyd's theories also incorporated many insights from cybernetics and systems thinking. Cybernetics, which informed the cognitive revolution, introduced the elements of *feedback* and *homeostasis,* and Bertalanffy's *General Systems Theory* brought Boyd the recognition that living systems are not the closed systems described by cybernetics, but *open systems.*[34] Bertalanffy defined an open system as "a system in exchange of matter with its environment, presenting import and export, building up and breaking down of its material components. . . . Closed systems are systems which are considered to be isolated from their environment."[35] Unlike closed systems, open systems maintain themselves far from equilibrium in a "steady state" characterized by continual flow and change. An open system may "actively" pass from a lower to a higher state of order; feedback mechanisms in such systems can reactively reach a state of higher organization owing to "learning," that is, information fed into the system.[36] Ilya Prigogine later refined and mathematically proved these insights, which form the basis for chaos and complexity theory.[37] *The Strategic Game of ? and ?*, in which Boyd distills the abstract essence of strategy, includes the idea that "Living systems are open systems; closed systems are non-living systems. Point: if we don't communicate with the outside world—to gain information for knowledge and understanding as well as matter and energy for sustenance—we die out to become non-discerning and uninteresting part of that world."[38] Indeed, when we do not maintain communication with the outside world, Boyd asserts, both Gödel and the Second Law will "kick in": "One cannot determine the character or nature of a system within itself. Moreover, attempts to do so lead to confusion and disorder."[39] In *Patterns of Conflict* Boyd also refers to the Second Law by stating that one of the key elements of victory consists of "Diminish own friction (or entropy) and magnify adversary friction (or entropy)."[40]

Nonlinear Sciences and Complex Adaptive Systems

By the end of the 1970s the first studies on nonlinearity appeared in the popular press, and by the beginning of the 1990s chaos and complexity became bestselling themes in popular science. The works Boyd was reading and rereading at the time of his death almost exclusively concern neo-Darwinism and the so-called "new sciences." They include Brian Goodwin's *How the Leopard Changed Its Spots,* which deals with the evolution of complexity; the book is marked with a sticky note reading "Dad's favorite."[41] This literature revived and integrated many of the concepts Boyd had previously encountered and provided him with a unified theory and indeed a metaphor for explaining and expressing the dynamics of human conflict. From these works Boyd discovered the remarkable feature that different nonlinear systems have inherently identical structures. Biological evolution, the behavior of organisms in ecological systems, the operation of the mammalian immune system, learning and thinking in animals, the behavior of investors in financial markets, political parties, and ant colonies have common characteristics: they are complex adaptive systems that consist of networks of "agents" acting in parallel. In a brain the agents are nerve cells; in ecologies the agents are species; in the global economy the agents are firms, individuals, or even nations.[42] Each agent exists in an environment produced by its interactions with the other agents in the system. Agents constantly act and react to what the other agents do, and therefore essentially nothing in their environment remains fixed.

Complex adaptive systems have many levels of organization, with agents at any one level serving as the "building blocks" for agents at a higher level. Cells form a tissue, a collection of tissues forms an organ, organisms form an ecosystem. Hierarchies contain intercommunicating layers: agents exchange information within given levels of the hierarchy, and different levels pass information among themselves as well. The complex system operates on a corresponding number of disparate time and space scales.

Complex adaptive systems anticipate the future. Through conditional action and anticipation they maintain and exhibit coherence during times of change. For this they employ internal models of the world (as in systems theory) and are therefore robust (or fit). They resist perturbation or invasion by other systems and try to adapt to novelty and change. The works of Gell-Mann showed Boyd that several levels of adaptation may occur in complex systems. *Direct adaptation* takes place as a result of the operation of a schema that is dominant at a particular time (as in a thermostat or cybernetic device). None of the behavior requires any change in the prevailing schema. The next level involves *changes in the schema,* competition among various schemata, and the promotion or demotion of schemata depending on the action of selection pressures in the real world.[43] The third level of adaptation is the Darwinian *survival of the fittest.* An organization or society may simply cease to exist if its schemata fail to cope with events.

The three levels of adaptation take place, generally speaking, on different time scales. An existing dominant schema can be translated into action right away, within days or months. A revolution in the hierarchy of schemata is generally associated with a longer time scale, although the culminating events may come swiftly. Extinctions of societies usually takes place over still longer intervals. Obviously, whether adaptation succeeds at any one level depends in large part on the rate of change of the environment in relation to the rate of adaptation of which an organism is capable, a theme close to Boyd's heart.

Importantly, complex systems operate in accordance with the second law of thermodynamics, exhibiting entropy over time unless replenished with energy. Nonlinearity can stabilize systems as well as destabilize them; it can drive open systems to crisis points where they will either bifurcate and self-organize again or enter a period of stochastic chaos (exhibiting erratic randomness).

Various authors in addition to Gell-Mann argued that models of organization based on living systems are naturally organic and adaptive.[44] For organizations to survive, they need to coevolve with their environment while maintaining internal stability. This requires variety, creativity, and learning communities. An organization must embody enough diversity to stimulate learning but not enough to overwhelm the legitimate system and cause anarchy. These studies maintain that, like ecosystems, organizations thrive when composed of heterogeneous parts. Practically speaking, organizations should possess a range of coupling patterns, from tight to loose, that provide "glue" and coherence among the subsystems. Loosely coupled structures allow an organization to adjust to environmental drift and react sluggishly to shocks, thus buying time for recovery. Moderate and tightly coupled structures prevent organizations from overreacting to environmental perturbation. Coupling patterns, then, allow organizations to maintain relative stability in most environments and protect the system even against severe shocks.[45]

Robust systems, like vibrant ecosystems, are characterized by rich patterns of tight, moderate, and loosely coupled linkages. Chains of interdependency branch in complicated patterns across nearly every actor in a broad network of interaction. Such complex patterns of interaction protect the organization against environmental shock by providing multiple paths for action. If one pattern of interdependency in a network is disrupted, the dynamic performed by that subsystem can usually be rerouted to other areas of the network. Such robustness makes it difficult to damage or destroy the complex system.[46]

Structurally, the key characteristic of complex adaptive systems is self-organization.[47] Proponents of this view see the "cellular" form operating in a network as the optimal organizational form for adaptation in turbulent environments, an idea included in Boyd's views on command and control. Small teams operating relatively autonomously pursue entrepreneurial opportunities and share knowledge and skills; meanwhile, the shared values of corporate culture in belief systems provide tight, but internal and perhaps even "tacit" control as a form of protocol.[48] At the same time, loose control comes from interaction between supervisor and employees that encourages information sharing, trust, and learning. The key to loose control is management's trust that employees will act according to the shared values, which sets them free to search for opportunities, learn, and apply their accumulating knowledge to innovative efforts.[49] Successful leaders of complex organizations allow experimentation, mistakes, contradictions, uncertainty, and paradox so the organization can evolve. Managing an organization as a complex system means relaxing control and focusing instead on the power of the interconnected world of relationships and the feedback loops.

These insights also show parallels with Japanese management practices that Boyd read about during the late 1980s and early 1990s and with historical case studies of military command such as, for instance, van Creveld's *Command in War.* Several of these ideas surface in Boyd's vision of command and control, which hinges on trust, implicit communication, open flow of information, and a shared view of the organizational purpose. Boyd would also emphasize variety and flexibility as other key traits of an organization. The concept of loose coupling is evident in Boyd's advice to implement a relaxed approach to command, allowing units and commanders sufficient latitude to respond to and shape their rapidly changing environment.

Military studies rapidly began to explore nonlinear sciences, and Boyd both read and contributed to early articles in which these ideas made the leap into military thought. For example, Pat Pentland's *Center of Gravity Analysis and Chaos Theory* asserts that crisis points can be precipitated by closing a system off from its environment and propelling it into equilibrium, eliminating feedback within the system, driving any of the dimensional dynamics to singularity by overloading or destroying it, and applying quantum amounts of broad external energy to the entire system.[50]

Boyd also read Alan Beyerchen's article "Clausewitz, Nonlinearity, and the Unpredictability of War," which emphasizes the relevance of the new sciences to the study of war. It argues that the cornerstone of Western military thought, Clausewitz's *On War,* implicitly incorporated core themes of "chaoplexity."[51] Beyerchen asserts that Clausewitz understood war as inherently a nonlinear phenomenon and perceived the nature of war as an energy-consuming phenomenon involving competing and interactive factors. Unpredictability, which results from interaction, friction, and chance, is a key manifestation of the role that nonlinearity plays in Clausewitz's work. A military action produces not a single reaction, but dynamic interactions and anticipations that pose a fundamental problem for any theory. Strategists can envision such patterns only in qualitative and general terms, not in the specific detail needed for prediction.

SNAPSHOTS OF *A DISCOURSE*

The previous section revealed that Boyd's ideas actually rest on a firm scientific foundation and do not merely reflect inductive thinking. The eclectic multidimensional and holistic approach of these theorists was essential to understanding the complex behavior of complex systems, and this mode of thinking became an argument in itself. Boyd wanted to inculcate his audience not so much with a doctrine as with an understanding of the dynamics of war and strategy and a style of thinking about that dynamic that differed from the deterministic mindset that prevailed in the strategic discourse of the 1960s and 1970s.

This theoretical framework also resulted in a unique set of terms and concepts—a new language—to express strategic behavior. Not only did Boyd argue that a multidisciplinary approach informed by insights from a variety of scientific fields is a prerequisite for sound strategic thinking, but that science also led him to new perspectives, hypotheses, and insights and helped him explain and connect ideas in a novel way. Thus he introduced into strategic theory the concept of open, complex, adaptive systems struggling to survive in a contested, dynamic, nonlinear world permeated by uncertainty. These systems constantly attempt to improve and update their schemata, their repertoire of actions, and their position in the ecology of the organization. Indeed, science provided him with novel metaphors, analogies, and illustrations—with novel conceptual lenses—to approach and explain military conflict. Boyd hints at this in the assertion that "the key statements of these presentations, the OODA Loop Sketch and related insights, represent an evolving, open-ended, far-from-equilibrium process of self-organization, emergence and natural selection."[52]

From Epistemology to Flying Fighters and Making Strategy

A brief examination of Boyd's work further illustrates this point and introduces several other ideas and lines of reasoning, showing that the OODA loop, and his overall work, have more substance than is often acknowledged. Boyd lays out the scope of *A Discourse* on page 1: "To flourish and grow in a many-sided uncertain and ever changing world that surrounds us, suggests that we have to make intuitive within ourselves those many practices we need to meet the exigencies of that world. The contents that comprise this 'Discourse' unfold observations and ideas that contribute towards achieving or thwarting such an aim or purpose."

The main vehicle Boyd uses to communicate his ideas is *Patterns of Conflict*. His departure point is the assumption that success in war, conflict, and competition—indeed, survival—hinges on the quality and tempo of the cognitive processes of leaders and their organizations and on their ability to translate these processes into relevant responses to a dynamic, ambiguous, and threatening environment. If conflict is a permanent feature of society, and adaptability is the key to survival, he suggests that "variety, rapidity, harmony and initiative seem to be the key qualities that permit one to shape and adapt to an ever-changing environment."[53]

Boyd derives this point of departure from the presentation *A New Conception for Air to Air Combat*, which he developed in 1975 while working on *Patterns of Conflict* and *Destruction and Creation* simultaneously. To refute those who accuse Boyd of an unwarranted inductive approach, it is useful to see how Boyd's thinking made the radical transition from air-to-air combat to operational art and strategic behavior in the wider sense. In *A New Conception for Air to Air Combat* he combines insights he had already gained during his research for *Destruction and Creation* with his knowledge of air-to-air combat, and he manages to describe air-to-air combat in terminology that relates to adaptability and cognitive processes.

In *A New Conception for Air to Air Combat* Boyd looks at maneuverability, a topic that came to the fore because in the fly-off competition the YF-16 had dramatically and unexpectedly outperformed the YF-18 even though energy maneuverability diagrams had predicted a close contest. The cause was transition rate, and subsequently Boyd formulates the suggestion that "in order to win or gain superiority—we should operate at a faster tempo than our adversaries or inside our adversaries time scales . . . such activity will make us appear ambiguous (non predictable) thereby generate confusion and disorder among our adversaries."[54] He adds that these suggestions are in accordance with "Gödel's Proof, The Heisenberg Principle and the Second Law of Thermodynamics," ideas central to *Destruction and Creation*.[55] Making the giant leap from air-to-air combat to warfare in general, he continues: "Fast transients (faster tempo) together with synthesis associated with Gödel, Heisenberg, and the Second Law suggest a New Conception for Air-to-Air Combat and for Waging War."[56]

In elaborating this new concept, Boyd stated that one's actions should "exploit operational and technical features to generate a rapidly changing environment (quick/clear observations, fast tempo, fast transients, quick kill)." Introducing key themes that also appeared in later presentations, Boyd further asserts that one should "inhibit an adversary's capacity to adapt to such an environment (suppress or distort observations)." The goal of such actions is to "unstructure adversaries system into a 'hodge podge' of confusion and disorder by causing him to over or under react because of activity that appears uncertain, ambiguous or chaotic."[57] The last slide contains Boyd's (Darwinistic) "message," one that similarly permeates his later work: "he who can handle the quickest rate of change survives."[58] This insight could apply to the dynamic of any combat at the tactical, operational, and strategic levels of war.

Thus the overarching theme throughout Boyd's theories concerning winning and losing is not just rapid decision making, but, as he states, "the . . . strategic aim [is] to diminish adversary's capacity to adapt while improving our capacity to adapt as an organic whole, so that our adversary cannot cope while we can cope with events/efforts as they unfold."[59] Boyd explicitly expresses the aim of any organism and organization in neo-Darwinistic and system-theoretical terms, and he regards the contestants—platoons, armies, their headquarters, businesses, and societies—as living systems, as complex adaptive systems. As he asserts in *Patterns of Conflict*, all such systems strive to "improve fitness as an organic whole, to shape and cope with an ever changing environment."[60]

Aiming for Organizational Collapse

Patterns of Conflict, the 193-slide opus that captures Boyd's research on conflict and warfare, illustrates, confirms, and elaborates upon these themes in a selective and biased survey of the evolution of warfare and of theories about gaining a victory. The survey does not cover specific areas such as air warfare, maritime strategy, and nuclear strategy, but it is certainly not limited in the comprehensiveness of its exploration of land warfare, thus catering to the needs of his primary audience: the U.S. Army and U.S. Marine Corps.

One could consider the introduction to briefing akin to a statement of various hypotheses, namely the assertion that if conflict is a permanent feature of society and adaptability is the key to survival, "variety, rapidity, harmony and initiative seem to be the key qualities that permit one to shape and adapt to an ever-changing environment." At the end of the presentation Boyd returns to these "key qualities," asserting that his analysis in *Patterns of Conflict* has confirmed their importance. In between lies a conscious effort to uncover the dynamics of combat in which he fleshes out the details of what it means to "get inside the OODA loop" of the opponent at the various levels of war.

A crucial and original section of *Patterns of Conflict* involves an exploration of three modes of conflict that Boyd distinguished in his panorama of military history:[61]

> *Attrition Warfare* as practiced by the Emperor Napoleon, by all sides during the 19th century and during World War I, by the Allies during World War II, and by present-day nuclear planners. *Maneuver Conflict* as practiced by the Mongols, General Bonaparte, Confederate General Stonewall Jackson, Union General Ulysses S. Grant, Hitler's Generals (in particular Manstein, Guderian, Balck, Rommel) and the Americans under Generals Patton and MacArthur. *Moral Conflict* as practiced by the Mongols, most Guerrilla Leaders, a very few Counter-Guerrillas (such as Magsaysay) and certain others from Sun Tzu to the present.

Contrary to what his critics assert, Boyd acknowledges the merits of attrition warfare in principle and is also keenly aware that blitzkrieg and guerrilla warfare do not guarantee success, as the history of warfare has shown that there are effective counters to both. *Patterns of Conflict* is replete with elaborate sections showing the dialectic process of move and countermove.

Boyd communicated the essence of attrition warfare in the slide reproduced as figure 2.2.[62] Boyd notes the drawbacks of attrition warfare, which are that it emphasizes mass, kinetic force, and thus the physical domain, and is also deterministic, linear, and bloody. Attrition warfare underutilizes leverage in the mental and moral domains; hence Boyd's preference for maneuver and moral conflict. Both concepts are based on an original and innovative synthesis of the essences of guerrilla warfare and blitzkrieg. In both styles, he argues, combatants avoid battles. Instead, the essence of both is to "*penetrate an adversary* to subvert, disrupt or seize those connections, centers, and activities that provide cohesion (e.g., psychological/moral bonds, communications, lines of communication, command and supply centers, . . .)[;] *exploit ambiguity,* deception, superior mobility and sudden violence to generate initial surprise and shock, again and again and again[; and] *roll-up/wipe-out,* the isolated units or remnants created by subversion, surprise, shock, disruption

FIGURE 2.2 ESSENCE OF ATTRITION WARFARE

Create and exploit

- **Destructive force:**
  Weapons (mechanical, chemical, biological, nuclear, etc.) that kill, maim, and/or otherwise generate widespread destruction.
- **Protection:**
  Ability to minimalize the concentrated and explosive expression of destructive force by taking cover behind natural or manmade obstacles, by dispersion of people and resources, and by being obscure using camouflage, smoke, etc., together with cover and dispersion.
- **Mobility:**
  Speed and rapidity to focus destructive force or move away from adversary's destructive focus.

Payoff

- Frightful and debilitating attrition via widespread destruction as basis to:
  - Break enemy's will to resist
  - Seize and hold terrain objectives

Aim

Compel enemy to surrender and sue for peace.

and seizure."[63] These actions aim to "exploit subversion, surprise, shock, disruption and seizure to generate confusion, disorder, panic, etc., thereby shatter cohesion, paralyze effort and bring about adversary collapse."[64]

The extraordinary level of success, or in Boyd's words "the message," results because in both approaches to conflict "one operates in a directed yet more indistinct, more irregular and quicker manner than one's adversaries (and thus not just fast)." This enables one to "Repeatedly concentrate or disperse more inconspicuously and/or more quickly from or to lower levels of distinction (operational, organizational and environmental) without losing internal harmony." For the same reason, one is able to "Repeatedly and unexpectedly infiltrate or penetrate adversaries' vulnerabilities and weaknesses in order to splinter, isolate or envelop and overwhelm disconnected remnants of adversary organism." Put in another way, one can "operate inside the enemy's OODA loops or get inside their mind-time-space as a basis to penetrate the moral-mental-physical being of one's adversaries in order to pull them apart and bring about their collapse."[65] Boyd notes that such "amorphous, lethal, and unpredictable activity by Blitz and Guerrillas make them appear awesome and unstoppable which altogether produce uncertainty, doubt, mistrust, confusion, disorder, fear, panic and ultimately collapse." It affects the connections and centers that provide cohesion, as Boyd explains in yet another slide on the same theme.[66]

These insights recur in the concept of moral and maneuver conflict. Moral conflict leverages mistrust and disinformation to sever the bonds within an organization, while maneuver conflict aims to create and exploit an information differential, thus affecting the opponent's capacity to adapt. Boyd's concept not only regards disorientation as the single element affecting adaptability, but also the element of overload due to "a welter of threatening events."[67]

Ambiguous information and uncertainty reduce adaptability, an effect compounded by the fear that results from threatening events. Interestingly, employing the various elements of maneuver conflict may not directly result in collapse. Instead, Boyd considers it equally valuable to try and create many isolated remnants of enemy forces that can later be mopped up. In light of the critique that Boyd expects victory through merely repeating OODA cycles more rapidly, this is not a trivial observation. Boyd described the "essence of maneuver conflict" in the terms seen in figure 2.3.[68]

FIGURE 2.3 ESSENCE OF MANEUVER CONFLICT

Create, exploit, and magnify	***Payoff***
• **Ambiguity:** Alternative or competing impressions of events as they may or may not be.	• **Disorientation:** Mismatch between events one (seemingly) observes or anticipates and events (or efforts) he must react or adapt to.
• **Deception:** An impression of events as they are not.	
• **Novelty:** Impressions associated with events/ideas that are unfamiliar or have not been experienced before.	• **Surprise:** Disorientation generated by perceiving extreme change (of events or efforts) over a short period of time.
• **Fast transient maneuvers:** Irregular and rapid/abrupt shift from one maneuver event/state to another.	• **Shock:** Paralyzing state of disorientation generated by extreme or violent change (of events or efforts) over a short period of time.
• **Effort (Cheng/Ch'I or Nebenpunkte/Schwerpunkt):** An expenditure of energy or an irruption of violence—focused into, or through, features that permit an organic whole to exist.	• **Disruption:** State of being split apart, broken up, or torn asunder.

Aim

Generate many non-cooperative centers of gravity, as well as to disorient or disrupt those that the adversary depends upon, in order to magnify friction, shatter cohesion, produce paralysis, and bring about his collapse.

In figure 2.4, which produces his slide on the "essence of moral conflict," Boyd includes ideas to counter mistrust, uncertainty, and danger, again demonstrating his emphasis on adaptability.

FIGURE 2.4 ESSENCE OF MORAL CONFLICT

Create, exploit, and magnify

- **Menace:**
  Impressions of danger to one's well-being and survival.
- **Uncertainty:**
  Impressions, or atmosphere, generated by events that appear ambiguous, erratic, contradictory, unfamiliar, chaotic, etc.
- **Mistrust:**
  Atmosphere of doubt and suspicion that loosens human bonds among members of an organic whole or between organic wholes.

Idea

- Surface, ***fear, anxiety***, and ***alienation*** in order to generate many non-cooperative centers of gravity, as well as subvert those that adversary depends upon, thereby magnify internal friction.

Aim

Destroy moral bonds that permit an organic whole to exist.

Synthesizing the essence of both modes of conflict, in the slide reproduced as figure 2.5 Boyd subsequently arrives at a plausible approach to induce what he calls "disintegration and collapse."[69]

Clearly, Boyd's ideas transcend the simple OODA loop, but that concept also has more meaning than is generally recognized. Boyd's reconceptualization of the meaning of tactics and strategy confirms this insight, as discussed next.

Redefining Tactics and Strategy

After examining the inner workings of blitzkrieg and guerrilla warfare, Boyd offers original reinterpretations of the dynamics at play at the four levels of war, expressing them in relation to the theme of adaptation. At the tactical and grand tactical (i.e., operational) levels, adaptation can be seen as a function of the speed of action and reaction and of information availability. Here a combatant can indeed achieve success by translating information into appropriate action faster than an adversary can, as in aerial combat. Even in this framework, however, the speed of decision making should be linked to a variety of actions, to irregularity and ambiguity, and to a rapid but varying pace, thus preventing predictability and forcing the opponent to react quickly.

FIGURE 2.5 THEME FOR DISINTEGRATION AND COLLAPSE

Synthesis

- **Lethal effort:**
  Tie up, divert, or drain away adversary attention and strength as well as (or thereby) overload critical vulnerabilities and generate weaknesses.
- **Maneuver:**
  Subvert, disorient, disrupt, overload, or seize those vulnerable yet critical connections, centers, and activities as basis to penetrate, splinter, and isolate remnants of adversary organism for mop up or absorption.
- **Moral:**
  Create an atmosphere of fear, anxiety, and alienation to sever human bonds that permit an organic whole to exist.

Idea

- Destroy adversary's moral-mental-physical harmony, produce paralysis, and collapse his will to resist.

Aim

Render adversary powerless by denying him the opportunity to cope with unfolding circumstances.

Boyd emphasizes a constant relationship between physical action (or the threat thereof) and its effects on morale and the enemy command process. Rapid, unexpected actions in quick succession have a great demoralizing effect and can disrupt tactical units' ability to interact. Even at these levels of combat, rapid OODA looping represents an oversimplified interpretation of the OODA loop concept. As Boyd explains in *Patterns:*

> ***Tactics***
> "OODA" more inconspicuously, more quickly and with more irregularity as basis to keep or gain initiative as well as shape or shift main effort; to repeatedly and unexpectedly penetrate vulnerabilities and weaknesses exposed by that effort or other effort(s) that tie up, divert, or drain-away adversary attention (and strength) elsewhere.
>
> ***Grand Tactics***
> Operate inside adversary's OODA-loops, or get inside his mind-time-space, to create a tangle of threatening and/or non-threatening events/efforts as well as repeatedly generate mismatches between those events/

efforts adversary observes, or anticipates, and those he must react to, to survive;

> ***Thereby***
> Enmesh adversary in an amorphous, menacing, and unpredictable world of uncertainty, doubt, mistrust, confusion, disorder, fear, panic, chaos, . . . and/or fold adversary back inside himself;
>
> ***Thereby***
> Maneuver adversary beyond his moral-mental-physical capacity to adapt or endure so that he can neither divine our intentions nor focus his efforts to cope with the unfolding strategic design or related decisive strokes as they penetrate, splinter, isolate or envelop, and overwhelm him.[70]

At the *strategic level,* Boyd notes, adaptation is more indirect and takes longer. Here the rapid OODA construct is clearly insufficient and less applicable. At this level social systems are characterized by a wide variety of subsystems and layers and thus a greater ability to counter enemy actions. Strategy is less about physical action and a rapid pace than about the ability to influence various dimensions of the opponent and limit its capacity to respond to actions that affect both its system and its environment. The game is to "Penetrate moral-mental-physical being to dissolve his moral fiber, disorient his mental images, disrupt his operations, and overload his system, as well as subvert, shatter, seize or otherwise subdue those moral-mental-physical bastions, connections, or activities that he depends upon, in order to destroy internal harmony, produce paralysis, and collapse adversary's will to resist."[71]

At the *grand-strategic level* (a level that receives a superficial and perhaps naïve treatment compared to the others) adaptability revolves around shaping the political and societal environment, which includes formulating an attractive ideology and adopting a mode of warfare the opponent is ill-suited to wage. Leaders should develop inspiring national goals and philosophies that unite and guide the nation as well as attract the uncommitted. Meanwhile, they should demonstrate that the opponent's government is corrupt, morally bankrupt, and disconnected from the population, and they should provoke the enemy to undertake actions that are viewed as disproportional and ineffective. Or, in Boyd's words, "Shape pursuit of national goal so that we not only amplify our spirit and strength (while undermining and isolating our adversaries) but also influence the uncommitted or potential adversaries so that they are drawn toward our philosophy and are empathetic toward our success."[72]

Returning to the hypothesis offered in the first slides, and marrying epistemology to blitzkrieg and guerrilla warfare, Boyd states that:

> He who is willing and able to take the initiative to exploit variety, rapidity, and harmony—as basis to create as well as adapt to the more

> indistinct—more irregular—quicker changes of rhythm and pattern, yet shape focus and direction of effort—survives and dominates. . . . The Game is to create tangles of threatening and/or non-threatening events/efforts as well as repeatedly generate mismatches between those events/efforts adversary observes or imagines (Cheng/Nebenpunkte) and those he must react to (Ch'i/Schwerpunkt) as basis to penetrate adversary organism to sever his moral bonds, disorient his mental images, disrupt his operations, and overload his system, as well as subvert, shatter, seize or otherwise subdue those moral-mental-physical bastions, connections, or activities that he depends upon thereby pull adversary apart, produce paralysis, and collapse his will to resist.[73]

In a strategic sense, he continues,

> [W]e need a variety of possibilities as well as the rapidity to implement and shift among them. The ability to simultaneously and sequentially generate many different possibilities as well as rapidly implement and shift among them permits one to repeatedly generate mismatches between events/efforts adversary observes or imagines and those he must respond to (to survive). Without a variety of possibilities adversary is given the opportunity to read as well as adapt to events and efforts as they unfold.[74]

Hence, Boyd asserts,

> Variety and rapidity allow one to magnify the adversary's friction, hence to stretch-out his time to respond. Harmony and initiative stand and work on the opposite side by diminishing one's own friction, hence compressing one's own time to exploit variety/rapidity in a directed way. Altogether variety/rapidity/ harmony/initiative enable one to operate inside adversary's observation-orientation-decision-action loops to enmesh adversary in a world of uncertainty, doubt, mistrust, confusion, disorder, fear, panic, chaos, . . . and/or fold adversary back inside himself so that he cannot cope with events/efforts as they unfold.[75]

In *The Strategic Game* he concludes that, at the most abstract level, strategy aims "to improve our ability to shape and adapt to unfolding circumstances, so that we (as individuals or as groups or as a culture or as a nation-state) can survive on our own terms."[76] Adapting insights from systems theory, complexity theory, and evolution theory, Boyd posits that strategy resembles a game of "interaction and isolation": "interaction permits vitality and growth while isolation leads to decay and disintegration."[77] In other words, it is "a game in which we must be able to diminish an adversary's ability to communicate or interact with his environment while sustaining or improving ours."[78] Isolate an opponent and in due course it will lose internal cohesion and external support, its delayed and misinformed reactions will be ineffective, and it will fail to adjust correctly to the changed environment. The aim is thus to change the opponent from an open to a closed system that slowly suffers the fate of all closed systems: the entropy that results from the second law of thermodynamics. The corollary is the imperative to maintain constant interaction among the units of one's own organization and between the organization and its environment.

Within the adversary system Boyd discerns physical, mental, and moral dimensions. Subsequently, isolation can occur—or be aimed for—in these different dimensions:

> Physical isolation occurs when we fail to gain support in the form of matter-energy-information from others outside ourselves. Mental isolation occurs when we fail to discern, perceive, or make sense out of what's going on around ourselves. Moral isolation occurs when we fail to abide by codes of conduct or standards of behavior in a manner deemed acceptable or essential by others outside ourselves.[79]

Interaction ensures the opposite:

> Physical interaction occurs when we freely exchange matter-energy-information with others outside ourselves. Mental interaction occurs when we generate images or impressions that match up with the events or happenings that unfold around ourselves. Moral interaction occurs when we live by the code of conduct or standards of behavior that we profess, and others expect us, to uphold.[80]

Physically, Boyd argues,

> [W]e can isolate our adversaries by severing their communications with the outside world as well as by severing their internal communications to one another. We can accomplish this by cutting them off from their allies and the uncommitted via diplomatic, psychological and other efforts. To cut them off from one another we should penetrate their system by being unpredictable, otherwise they can counter our efforts. Mentally we can isolate our adversaries by presenting them with ambiguous, deceptive or novel situations, as well as by operating at a tempo or rhythm they can neither make out nor keep up with. Operating inside their OODA loops will accomplish just this by disorienting or twisting their mental images so that they can neither appreciate nor cope with what's really going on. Morally our adversaries isolate themselves [!] when they visibly improve their well being to the detriment of others (allies, the uncommitted), by violating codes of conduct or behavior patterns that they profess to uphold or others expect them to uphold.[81]

Creating Agile Organizations

Boyd's advice for organizational culture, structure, leadership, and communication processes, elaborated in *Organic Design for Command and Control* and *The Conceptual Spiral,* is consistent with this argument. The key challenge is to maintain cohesion while conducting fluid, varied, and rapid actions despite uncertainty and threats. This solution stems from *Destruction and Creation* and his study of blitzkrieg and guerrilla warfare, but also from organization theory, theories on organizational and individual learning, and studies such as van Creveld's *Command in War* as well as recent works on organizational resilience and adaptability.[82] First, Boyd advocates a multispectral orientation: "We must interact in a variety of ways with our environment, and examine the world from a number of perspectives so that we can generate mental images or impressions that correspond to that world."[83] As he notes in *The Strategic Game,* this can only be accomplished by a continuous process of "analysis/synthesis across a variety of domains" that would produce a repertoire of "Orientation patterns."[84] "Orientation is the Schwerpunkt," he argues in *Organic Design.* "[I]t shapes the way we interact with the environment, the way we observe, decide and act."[85]

The answer to uncertainty lies in the degree of autonomy of action at the tactical level of an organization. In his presentation *Organic Design for Command and Control,* Boyd advocates creating adaptable, open learning organizations consisting of people with varied backgrounds cooperating in informally networked teams that can comfortably operate in an insecure environment because they have relatively low information requirements. A high degree of mutual trust and shared orientation patterns created by common experience and training, combined with clear objectives and common doctrine, ensures that a minimum of communication suffices to maintain coherence among the actions of various units. In contrast to standard military hierarchical cultures with top-down control, each level and each unit must have the autonomy and repertoire of resources it needs to enable individual initiative and creativity. These, in turn, are driven by the need to deal with the degree of dynamism and uncertainty of the immediate organizational environment.

Moreover, contrary to the opinions of critics who fault Boyd for proposing a single OODA loop, Boyd fully recognized the interaction of multiple and layered OODA loops in the various levels of the military organization. As he explains in *Patterns of Conflict,* for instance, blitzkrieg employs a concept of command and control in which each unit at the different levels of organization, from simple to complex, has its own OODA time cycle. The cycle-time increases at the higher organizational levels, as a unit tries to control more levels and issues. As the number of events increases, it takes longer to observe, orient, decide, and act. Thus "the faster rhythm of the lower levels must work within the larger and slower rhythm of the higher levels so that overall system does not lose its cohesion or coherency."[86]

Boyd indicates that in such an organization, both "command" and "control" are inappropriate terms. Higher command levels must restrain their desire to know about everything that occurs at lower levels and to interfere. They must encourage cooperation and consultation among lower levels and remain open to suggestions, bottom-up initiatives, and critique. Thus higher levels must exercise "leadership and appreciation" of activities and compare actions and events to their expectations. This approach to organization and leadership would allow a high tempo of operations and exploitation of opportunities. It would limit the need for time-consuming, detailed, top-down guidance and hence facilitate rapid adaptation to changing circumstances. This idea, which echoes the literature on Japanese management techniques and complex adaptive systems, also surfaces in *The Conceptual Spiral*, where Boyd argues that successful organizations survive, and actually thrive, because of their ability to continually innovate and experiment to deal with uncertainty and to introduce novelty into their environment. Interestingly, Boyd's arguments resonate closely with recent works on organizational resilience and adaptability.[87]

The Real OODA Loop

In *The Essence of Winning and Losing*, Boyd interweaves and condenses these arguments. Figure 2.6 depicts an elaborate, double-loop cybernetic model, informed by models found in academic articles and books on cognitive science and systems theory and in works on complex and chaotic systems. At its heart the OODA loop is much less a model of decision making than an epistemological model influenced by scientists such as Heisenberg, Gödel, Popper, Bronowski, Kuhn, and Polanyi, anthropologists such as Geertz, information theorists and cyberneticists such as Wiener and Neumann, system theorists such as Bertalanffy, and a host of others, including Darwin. It is a model of individual- and organizational-level learning and adaptation processes, or—to use Boyd's own terms—a meta-paradigm of mind and universe, a dialectic engine, an inductive-deductive engine of progress, a paradigm for survival and growth, and a theory of intellectual evolution.[88] As noted previously, his academic zeitgeist and its influence on his thinking come explicitly to the fore when he observes that "the OODA loop sketch and related insights represent an evolving, open-ended, far from equilibrium process of self-organization, emergence and natural selection."[89]

This illustration is preceded by five (postmodern) "key statements" that demonstrate the interrelationships among the core themes and insights in his various presentations:

> Without our genetic heritage, cultural traditions, and previous experiences, we do not possess an implicit repertoire of psychological skills shaped by environments and changes that have been previously experienced.
>
> Without analysis and synthesis, across a variety of domains or across a variety of competing/independent channels of information, we cannot evolve new repertoires to deal with unfamiliar phenomena or unforeseen change.

FIGURE 2.6 THE REAL OODA LOOP

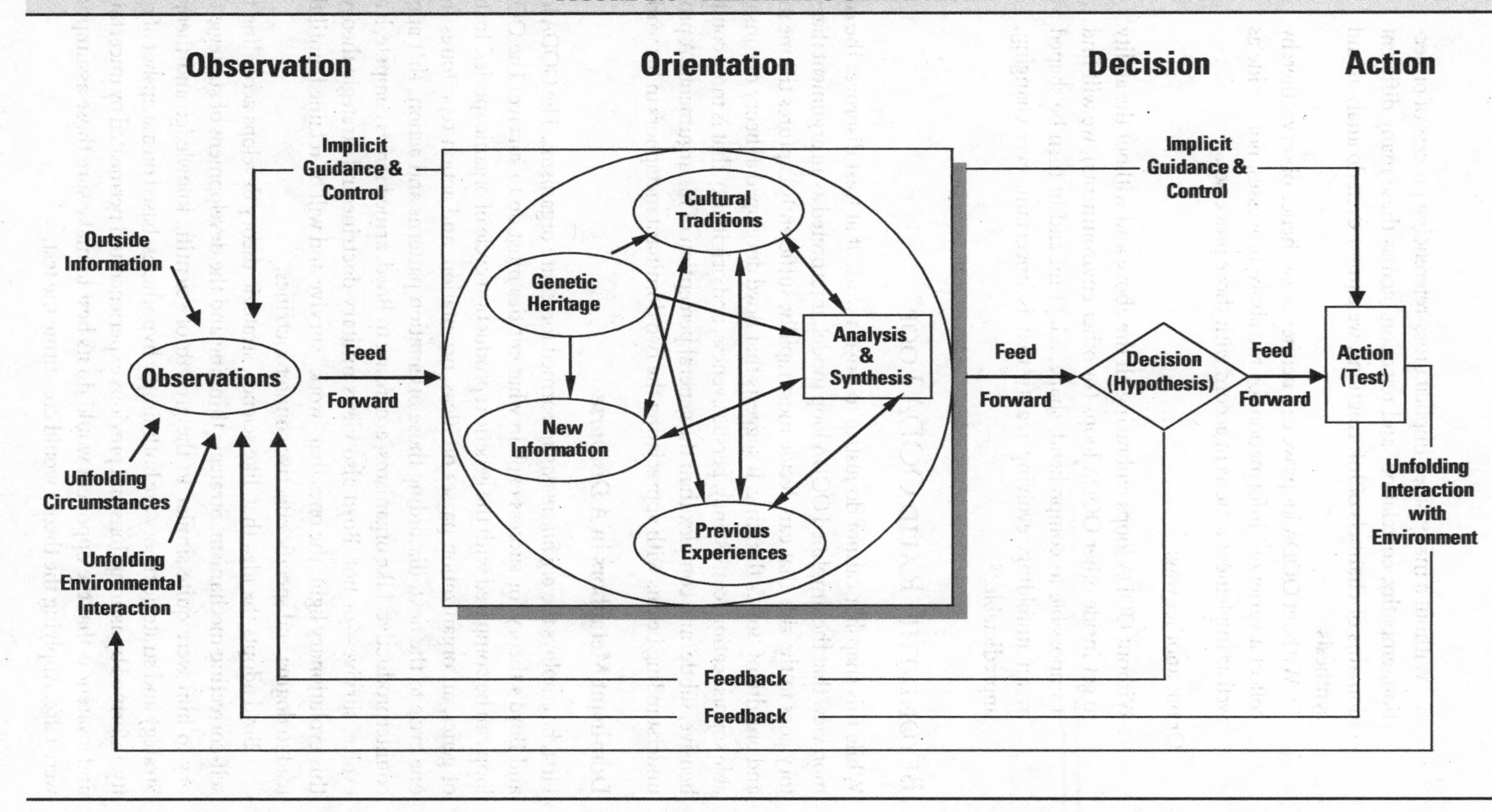

> Without a many-sided implicit cross-referencing process of projection, empathy, correlation, and rejection (across these many different domains or channels of information), we cannot even do analysis and synthesis.
>
> Without OODA loops we can neither sense, hence observe, thereby collect a variety of information for the above processes, nor decide as well as implement actions in accord with those processes.

Or put another way:

> Without OODA loops embracing all the above and without the ability to get inside other OODA loops (or other environments), we will find it impossible to comprehend, shape, adapt to, and in turn be shaped by an unfolding, evolving reality that is uncertain, ever-changing, unpredictable.[90]

Beyond the "Rapid OODA Loop"

While this snapshot cannot do justice to Boyd's work, it at least disproves the common view that the simplified OODA loop model, interpreted as an argument that victory goes to the side that can decide most rapidly, sufficiently captures the meaning and breadth of Boyd's thinking. It suggests that Boyd developed a theory of considerable sophistication, consistency, persuasiveness, and originality that is more comprehensive, subtle, and complex than the general perception of his arguments. A proper understanding begins with appreciating the two dominant metaphors in his work.[91]

Dominant Metaphors in *A Discourse*

First, he employs the *organic metaphor:* armed forces are organisms. The OODA loop and Boyd's theory for success explain what entities must do to survive. The OODA loop can be compared with the genetic reproduction cycle of organic species. Instead of genes, an organization passes on ideas, orientation, and action repertoires from one cycle to the next, discarding those orientation patterns and actions that appear counterproductive. Like organisms, according to Boyd, armed forces compete, learn, evolve, survive—or not. Boyd also views military doctrine and strategic theory in this evolutionary light: the ones that "work" survive and will be retained, until they lead to dogma and specific effective counterdoctrines.

Boyd adopts the idea that, like science, strategic theory develops according to a self-correcting mechanism. Strategic thinking and the development of strategic theory to him were quite similar to the growth of scientific knowledge and theories. Strategy and strategic theory are both simply hypotheses, based on a snapshot of reality, informed by assumptions and previous experience and permeated by uncertainty and awareness that the opponent would do its best to make sure those assumptions were false. Applying the theory would constitute the test.

The second metaphor lies in considering *armed forces as open systems*. Throughout his presentations Boyd emphasizes connections and relationships among various interdependent subsystems laterally and among hierarchical levels of organization, and the need for continuous interaction with the environment. Extending this metaphor, Boyd also regards *armed forces as complex adaptive systems* such as ecosystems. Here the themes of nonlinearity, novelty, variety, levels of organized complexity, and different ways, modes, and timescales of adaptation and evolution come into play. War resembles the nonlinear clash of two complex adaptive systems. Units fill niches in the ecosystem and should operate with sufficient autonomy, yet interdependently with other systems. Doctrines, procedures, tactics, and organizational culture are *like* mental models that provide coherent behavior. They are also *like* schemata that must evolve. Armed forces are *like* autopoietic systems, continually making efforts to maintain their distinctive character despite the turbulent environment. Defeat, demoralization, and a disorganized retreat can be seen as symptoms of a disintegrating system, unable to cope in the available time with the radical changes imposed on it by the environment and unable to adjust its schemata or to speed up its rate of adjustment.

This has implications for the interpretation of Boyd's work, which revolves around the themes of organizational survival and its solution: adaptability. Boyd focuses on the factors that can impair an opponent's ability to adapt and those that preserve one's own capacity to do so. The system for which the strategic theory is designed consists of armed forces and their environment. An armed force is by design a fairly robust system, structured to cause change within an opponent's system and resist external pressure to change itself. It equips itself with redundant connections, ample numbers of diverse units, good sensors, relevant schemata, and a supportive environment. It does anything necessary to ensure a modicum of coherence among its actions.

Military force therefore aims to push an opposing system away from its ordered, disciplined state toward one where the several subsystems must self-organize because they lack higher direction, and then toward a state of randomness, but not necessarily in such a time-sequenced order. Randomness—the loss of cohesion—is the opposite of the capability to adapt. Thus, the units may still exist, but not as part of a higher order complex system. This mechanism of decreasing cohesion and degrading the ability to adapt can be applied to any system and subsystem, down to the lowest level that can be described as a system: in armed forces, the individual soldier. But the goal is not necessarily to make the system completely disintegrate in one massive blow. Because feedback loops will cause inappropriate or slow responses to magnify in impact over time, the force need only create an initial advantage and prevent the opponent from compensating for it. The nature of a system can change from less complex to complex or, alternatively, to the point where almost no connections remain, where no information is shared, and where cohesion of action ceases to exist. At that point the different elements of a system act randomly and no longer constitute a part of a system.

More Than Speed

The OODA loop is undoubtedly an important element of Boyd's work. As an abstract model, it can serve as a framework to explain his thoughts. It runs through all the levels of military operations and aggregation levels of social systems. Nevertheless, the comprehensive overview of Boyd's work shows that the OODA loop represents and means more than a decision process, and in Boyd's hands the model contains elements for victory that go beyond information superiority and speed.

The first misconception about the OODA loop concerns speed. The rapid OODA looping idea suggests a focus on speed of decision making, in other words, "outlooping" the opponent by going through consecutive OODA cycles faster. Indeed, Boyd frequently suggested as much. But he also addressed the aspect of altering the tempo, making it difficult for the opponent to adapt adequately to the fast *changes* in the situation, including the element of speed. Thus it is not absolute speed that counts. It is the relative tempo or variation in rhythm that counts.

Second, and related to the previous observation, rapid OODA looping should not be equated with merely going through the decision cycle faster than the opponent and/or accomplishing this process with more information. Such a view ignores the close interrelationship between physical action and the mental and moral components. One party can have a distinct advantage in timely and accurate information, but this "information superiority" is useless if it cannot be translated into meaningful action. Boyd instead argues that strategy should aim to create and perpetuate a highly fluid and menacing state of affairs for the enemy and to disrupt or incapacitate the enemy's ability to adapt to such an environment. Thus the psychological (mental/moral) and temporal mechanisms come into play only if and when the physical and spatial dimensions are also adequately manipulated. Physically threatening and lethal actions interact with mental and moral effects. Although perception, and thus the psychological dimension, lies at the heart of the theory, this does not mean the physical aspects are less important. Boyd shows the close relationship of the two dimensions and highlights the effectiveness of the synthesis of actions in the physical, mental, and moral spheres.

Third, while tempo is important, it is just one of many other control dimensions in this theme. It is probably particularly important at the tactical level as a factor directly influencing chances of success. At the grand tactical level Boyd is more concerned with "operating inside adversary's mind-time-space" and with "generating mismatches between those events he observes, or anticipates and those he must react to." At this level, as well as at the strategic and the grand strategic levels, a force can manipulate other dimensions.

This concept also surfaces when Boyd discusses the three categories of conflicts. Negating time and information is perhaps a necessary mechanism that leads to inadequate reactions, but it is not sufficient. Other factors must combine with lack of time to induce the moral, mental, and physical inability to react. Physical as well as nonphysical connections must be severed through the game of isolation and interaction.

This requires irregular, inconspicuous, and varied actions, and it is these, combined with the menacing aspect of those actions and the impressions, that produce paralysis, disintegration, and noncooperative centers of gravity and that shatter cohesion. Disintegration and collapse render the adversary powerless by denying it the opportunity to cope with unfolding circumstances. Thus the temporal dimension plays a large role, but a role within the context of the organism's ability to adapt.

Adaptability versus Organizational Collapse

Maintaining the ability to adapt while denying it to the opponent is the all-embracing theme that connects the various parts of *A Discourse.* In *Patterns* Boyd notes that "Adaptability is the power to adjust or change in order to cope with new and unforeseen circumstances," and that "in dealing with uncertainty it seems to be the right counterweight." Thus the dominant and overarching theme is not the narrow interpretation of rapid OODA looping, or "decision superiority," but rather the ability to adapt to unfolding, multidimensional events that occur at different time scales. The "rapid OODA looping idea" in the narrow sense fits into the larger theme of adaptation, and even this narrow view gains meaning when expressed within the context of organizational adaptation.

Boyd argues that adaptation occurs across various time scales and develops a view of what adaptability means and requires at each level. Each level knows its specific "name of the game." At the tactical and operational levels actions, movements, attacks, feints, threats, etc., disrupt the enemy's organizational processes, confuse commanders and personnel, attrit its forces, and dislocate its units. Confusion, fear, and lack of information—or the inability to react to accurate perceptions of the threats—degrade trust, cohesion, and courage and thus the ability to cohere and respond collectively and to take the initiative, to adapt adequately as an organization. In this context, adaptation is direct.

At the *strategic level* adaptation is more indirect and takes longer. It revolves around adjusting doctrines and force structures and disrupting the opponent's orientation patterns or mental images. At the *grand strategic level* adaptation involves shaping the political and societal environment, including an attractive ideology, and selecting a form of warfare. Vitality and growth improve the fitness of an organic whole, creating the ability to shape and expand influence or power over events in the world. This also surfaces as the national goal, which emphasizes the effective combination of isolation and preservation strategies in all dimensions, the mental-moral as well as the physical. At this level Boyd disregards the temporal aspect, because success not only involves overloading the opponent's OODA system but also derives from the interplay of leveraging across multiple dimensions. Success results from playing the game of interaction and isolation well.

The Centrality of Orientation

The narrow interpretation of the OODA loop also overlooks another essential feature of Boyd's theory: *developing, maintaining, and reshaping one's orientation*, the box around which the loop graphically revolves (see Figure 2.6). Speed, or rather tempo, has little value if one cannot react adequately to incoming information or if one's interpretation of events is flawed. Brave decisions and heroic actions are pointless if inadequate orientation led to inaccurate observation. Orientation stems from genetic heritage, cultural tradition, experience, and unfolding circumstances and is shaped by the interplay of these factors. It is the "genetic code" of an organism or organization. For any command concept, then, orientation is the "Schwerpunkt."

To avoid predictability and ensure adaptability to a variety of challenges, having a single, common orientation, thought pattern, belief system, or military doctrine does not suffice to solve all operational contingencies. An entity must have a repertoire of orientation patterns and the ability to select the appropriate one for the situation at hand while denying the opponent the same capability. To maintain variety in response one should build variety into the construction of orientation patterns. Apart from combining an analytic with a synthetic approach, organizations accomplish this by involving people with varying backgrounds and experience and confronting the entire group with varying situations. This builds trust and variety in response and fosters communication about different people's ways of viewing situations. Boyd is highly aware of the need to exercise great care when selecting people for command and for operational headquarters. This concept, in turn, requires a common outlook and doctrine; otherwise units may respond in totally unexpected ways.

A final aspect Boyd stresses in relation to orientation is the importance of having a repertoire of relevant schemata *combined with* an ability to validate the schemata before and during operations and to devise and incorporate new ones, if one is to survive in a rapidly changing environment. Indeed, learning is essential for adaptation. One may react quickly to unfolding events, but if one is still constantly surprised one has apparently been unable to turn the findings of repeated observations and actions into a better appreciation of the opponent. In other words, one has not learned, but instead has continued to operate according to existing orientation patterns. Examining established beliefs and expectations and, if necessary, modifying them in a timely manner is crucial. As Boyd asserted in *The Strategic Game*, the way to play the game of interaction and isolation is to spontaneously generate new mental images that match an unfolding world of uncertainty and change.

A Discourse and Airpower: Contributing to Organizational Collapse

The Enemy as a Complex Adaptive System

A Discourse is not a "Theory of Everything," nor is it an airpower theory. Instead it is what Colin Gray calls a partial theory, with some elements of a general theory.[92] It has little to say about current security problems such as fragile states, the proliferation of WMD, nuclear strategy, maritime or air warfare (indeed, *Patterns of Conflict* comes down almost entirely to an analysis of land warfare), nor did Boyd intend it to do so. Yet it contains insights that have applicability beyond military conflicts and interstate warfare. It is strongest as an exploration of operational art, but also offers novel, highly conceptual, generic interpretations of the nature of tactics and strategy.

Boyd's work does not result in specific targeting suggestions, like those in Douhet, Mitchell, or John Warden, nor does it advocate the strategic use of airpower over tactical employment or inform coercive strategies for operations such as Allied Force over Kosovo. Importantly, he has nothing to say about the shifts in Western attitudes toward the legitimacy of using force. Yet although Boyd does not address airpower as such, as one might expect from a general theory of war, his work can be productively applied (with some reservations) to explain the utility and indeed the logic of airpower. It sheds light on the system-wide dynamics that airpower can set in motion and the effects it may aim to achieve.

The themes of adaptability and organizational collapse and the insights from nonlinear sciences provide the keys for relating Boyd's work to airpower theory. David Fadok, who conducted such an exercise, has argued that Boyd, like John Warden, aims for strategic paralysis.[93] This is true, but, as indicated, Boyd defined the effects differently, and while paralysis would be a desirable end result, he suggested that less far-reaching effects would accumulate over time and produce success in themselves or provide favorable conditions for further actions. His slides include references to various stages of degradation in organization effectiveness; as noted previously, the game is "to penetrate adversary organism to sever his moral bonds, disorient his mental images, disrupt his operations, and overload his system, as well as subvert, shatter, seize or otherwise subdue those moral-mental-physical bastions, connections, or activities that he depends upon thereby pull adversary apart, produce paralysis, and collapse his will to resist."[94]

Importantly, in contrast to airpower theorists such as Douhet, Mitchell, and perhaps Warden, all of whom worked on the premise that the enemy is a fragile system and described that system with metaphors such as "house of cards" or "spiderweb," Boyd assumes that the opponent is actually a robust system. Indeed, the starting point for linking Boyd's work and airpower is the image of the contestants as open adaptive systems composed of many interconnected layers and subsystems, with the game being to drive the opponent toward a closed condition, fully aware that the opponent will try to do likewise.

Given this conceptualization of war and armed forces as complex adaptive systems, social systems can employ the following set of generic strategic moves to influence other systems and accomplish their aims during a conflict. (1) Use multiple strategies/avenues. (2) Affect the accuracy of the cognitive/feedback process. If comprehension helps to achieve cohesion and maintain purposeful behavior, the corollary is that confusion helps to create collapse. (3) Overload the opponent's cognitive capacity. (4) Eliminate and/or threaten particular crucial (real or imaginary) subsystems. (5) Diminish the variety of subsystems (affecting the ability to respond to a variety of threats and diminishing the decision or adaptation space) by alternatively achieving and maintaining a relative and relevant advantage in variety. (6) Disrupt moral, physical, and/or informational vertical and horizontal relations (i.e., the interdependencies) and thus erode cohesion among subsystems. (7) Close the enemy off from its physical/social environment. (8) Shape the environment of a system faster than the opponent's capability to cope with it. (9) Disrupt the information flow between the environment and the system. (10) Ensure the irrelevance of the opponent's schemata, or make the enemy unable to validate those schemata while ensuring sufficient accuracy of one's own schemata. (11) Change the nature of war; wage a form of warfare that does not correspond to the opponent's doctrine and strategic preference (schemata). (12) Change the environment in terms of alliances, location, and/or stakes involved.

Boyd mentions all of them in his work. As a brief examination of recent joint campaigns—Desert Storm, Enduring Freedom, and Iraqi Freedom in particular—will suggest, airpower has great potential to perform most, if not all, of these tasks.

Airpower in the Postmodern Era

Airpower is a dominant feature of postmodern warfare, with precision sensors and weapons as the iconic symbols.[95] Stealth, electronic warfare capabilities, high-resolution sensors, and precision weapons have enabled airpower to become the leading edge of any substantial military operation involving the threat or actual use of force. The demonstrated ability to limit the level of destruction, civilian casualties, and collateral damage, and to minimize the risk to ground troops, have made the employment of airpower in lieu of and in support of ground operations a normative element. Ground operations now often support airpower, following the so-called "Afghan model," in which special forces, along with local proxy forces, operate as the eyes and target designators for air strike platforms.

According to many analysts, Desert Storm heralded this new "postmodern" age of warfare.[96] On the one hand it featured familiar NATO tactical air assets such as the Airborne Warning and Control System, specialized suppression of enemy air defenses aircraft, air-to-air refueling aircraft, F-16s, F-111s, and F-15 fighter-bombers. The opening phase of the liberation of Kuwait in January 1991 also seemed familiar, following NATO doctrine with offensive counter air operations to degrade the extensive Soviet-style Iraqi air defense system to gain air superiority. Moreover, the campaign included traditional air interdiction operations against Iraqi armor, logistics, and infrastructure, as well as close air support against dug-in Iraqi troops. On the other hand, the thirty-nine days of the massive, yet precise, air campaign, including conventional strategic attacks against targets in downtown Baghdad, that preceded the four-day ground campaign represented a break with the common and expected pattern of operations.[97]

Two icons stood out in creating this image.[98] The first was the advancement in precision in detection, identification, and attack capabilities. Precision-guided missiles led to a dramatic 13-to-1 differential between precision and nonprecision attacks, better than an order of magnitude.[99] Desert Storm also suggested that military operations need not entail massive civilian casualties and that the amount of "collateral damage" to civilian infrastructure seemed controllable. In addition, the risk to coalition troops was lower than expected. Approximately 148 coalition military personnel died in combat, a regrettable but also unprecedentedly low number considering the scale of the operation and the pessimistic prewar estimates of 10,000 coalition casualties.[100] The age of mass warfare—industrial-age warfare—that had existed since World War I was drawing to an end.

Stealth was the second icon. With a radar reflection surface the same size as that of a golf ball, the F-117s could operate almost unseen deep within enemy territory from the first moment of the war, sometimes attacking two targets per mission in the Baghdad area, which boasted the highest density of air defense systems in the world. Not only did stealth have immense operational value, but it also produced an indispensable improvement in efficiency. Whereas a typical nonstealth attack package required thirty-eight aircraft to enable eight of them to deliver bombs on three targets, only twenty F-117s were required to simultaneously attack thirty-seven targets successfully in the face of a strong air defense threat. The result was a dramatic rise in the intensity of attack. To illustrate, during 1943 Allied bombers attacked 123 target complexes in Germany; during the first twenty-four hours of Desert Storm coalition bombers attacked 148 target complexes.

The new dominance of offense over defense in air warfare as a result of stealth, standoff weapons, electronic warfare, and drones provided a sanctuary that could be exploited for various purposes. This represented a radical break with the past, when air warfare involved a continuous battle of attrition for air superiority that could mostly be obtained and exploited only over a limited area and for a brief period. The coalition had learned this lesson from, for instance, the Israeli wars of 1967 and 1973. In the Six-Day War of 1967, Israel disabled the Egyptian air force by attacking Egyptian airfields with impunity, achieving air superiority that proved instrumental in the success in the two-front war Israel had to fight on the ground. In contrast, the Yom Kippur War of 1973 demonstrated the increasing lethality of the air defense systems introduced in the 1960s. Egypt operated novel mobile Soviet SA-6 systems, resulting in devastating losses to the Israeli Air Force, giving the defense the upper hand. Desert Storm, however, suggested that even nonstealthy aircraft, supported by electronic warfare assets and equipped with precision munitions *and* precision information, could steer clear of advanced air defense systems by flying at high altitude while maintaining accuracy of attacks. The campaign thus demonstrated that advanced airpower capabilities offered the option to open a flank in the third dimension: a virtual sanctuary.

Intense day and night air attacks with precision-guided missiles offered the possibility of relatively quick success against traditional armed forces that relied on massed mechanized ground combat. In addition, relentless precision-guided missile attacks on trucks and bridges effectively halted the flow of supplies to the Iraqi front line and the movement of units within the Kuwaiti theater. Advanced aircraft proved devastatingly able to spot, identify, and destroy enemy troops on the move, whether in tanks or in transport convoys, effectively neutralizing enemy resistance before the troops could reach Western troops and making any substantial movement virtually impossible, either through physical destruction or through its demoralizing effect.[101]

Conservative postwar analysis indicates the terrible effectiveness of these attacks on tanks, mechanized vehicles, and artillery. In February 1991 almost two entire Iraqi divisions were destroyed from the air during their advance to Al Kafji after being detected by a Joint Surveillance Target Attack Radar System air-ground surveillance aircraft. Brigades suffered greater losses in half an hour of air attacks than during eight years of war against Iran.[102] As a result the Iraqi army chose to hunker down, and it continued to suffer mounting punishment, both physical and psychological, from the air. In the Kuwaiti theater, coalition air attacks managed to destroy sometimes in excess of 50 percent of Iraqi armor and artillery equipment, and Iraqi ground troops surrendered by the thousands after being pounded by B-52 strikes or showered with leaflets threatening such attacks. Only 20 percent of Iraqi tanks offered visible resistance after these attacks, and about 40 percent of Iraqi ground troops deserted before the actual coalition ground offensive.

Once again this suggested a break with experience. Interdiction and close air support had always been considered risky missions that required vast numbers of aircraft and had only limited chances of achieving significant effects. Now one fighter could attack several targets in one mission. The result was a drastic shortening of the time and risk involved for ground units to complete the coalition victory. It rendered archaic the traditional notion of massing a large ground force to confront an opponent, particularly on a "field of battle."[103]

The overwhelming airpower capabilities offered the potential to strike at the heart of a country (the regime) from the first moment of a campaign and cripple strategic command capabilities before attacking fielded forces.[104] Instead of focusing on armed forces exclusively, modern airpower made additional options available, such as regime targeting and precise disruption of infrastructure, with direct strategically significant effects. It also held out the promise that several smaller-scale physical and nonphysical attacks—with less destructive power—conducted simultaneously on vital and interdependent targets could achieve disproportionate strategic outcomes.

Desert Storm highlighted the advances in airpower's counter to land effectiveness and its value for joint warfare. It showed that airborne sensor platforms and command and control assets could provide unprecedented levels of situational awareness to the ground force commander. Airpower could now provide air superiority on a theater-wide scale, thus protecting ground forces, lines of supply, and logistics sites. It could prevent an adversary from massing armored forces, and it could delay, disrupt, and destroy follow-on forces, and isolate units. It could effectively threaten any governmental and economic infrastructure and facility, while reducing the risk of civilian casualties and collateral damage. It could operate at a higher intensity and could thus speed up the campaign tempo.

Operation Enduring Freedom

U.S. defense policy was quick to exploit these airpower capabilities. *Joint Vision 2010* condensed the tenets of postmodern warfare and U.S. defense aspirations in the following sentences, which directly relate to Boyd's work:

> By 2010 . . . instead of relying on massed forces and sequential operations, we will achieve massed effects in other ways. Information superiority and advances in technology will enable us to achieve the desired effects through the tailored application of joint combat power. . . . With precision targetting and longer range systems commanders can achieve the necessary destruction or suppression of enemy forces with fewer systems, thereby reducing the need for time consuming and risky massing of people and equipment. Improved command and control, based on fused, all-source real-time intelligence will reduce the need to assemble maneuver formations days and hours in advance of attacks. Providing improved targetting information directly to the most effective weapon system will potentially reduce the traditional force requirements at the point of main effort. All of this suggests that we will be increasingly able to accomplish the effects of mass—the necessary concentration of combat power at the decisive time and place—with less need to mass forces physically than in the past.[105]

In the latter part of the 1990s, the U.S. Department of Defense proposed an overarching concept: network-centric warfare. It became the foundational concept for the U.S. armed forces in the first decade of the twenty-first century, and with the agreement by NATO in 2002 to create the NATO Response Force and embark on "military transformation" it also entered the debates among European militaries. Network-centric warfare consists of a "set of warfighting concepts designed to create and leverage information."[106] The following description captures the essence: "[T]he US is poised to harness key information technologies—microelectronics, data networking, and software programming—to create a networked force, using weapons capable of pinpoint accuracy, launched from platforms beyond range of enemy weapons, utilizing the integrated data from all-seeing sensors, managed by intelligent command nodes. By distributing its forces, while still being able to concentrate fires, the US military is improving its mobility, speed, potency, and invulnerability to enemy attack."[107]

With explicit reference to Boyd's OODA loop, various Defense Department publications noted that the advantage for forces that implement network-centric warfare lies in gaining and exploiting an information edge.[108] These capabilities would allow a force to attain an improved information position that can partially "lift the fog of war" and enable commanders to improve their decision making and fight in ways that were not previously possible. Networking would facilitate the ability "to develop and share high quality situational awareness" and the "capability to develop a shared knowledge of commander's intent." This would enable "the capability to self-synchronize its operations." A force with these attributes and capabilities would be able to increase combat power by better synchronizing effects in the battlespace, achieving greater speed of command, and increasing lethality, survivability, and responsiveness.[109] The guidance in these documents corresponds well to the overall tenets put forward by Boyd, and it often refers explicitly to his work.

The toppling of the Taliban and the expulsion of Al-Qaeda from Afghanistan benefited from the developments of the 1990s. In particular, investments in data links and sensors (under the rubric of network-centric warfare and the *Joint Vision 2010*) had improved the capacity to observe, track, identify, and if necessary engage small mobile targets, a capability that had eluded the United States during Desert Storm and Allied Force, resulting in fruitless searches for Scud and mobile surface-to-air missile systems. With only three to five hundred special forces actually within Afghan territory, who in turn united and empowered local opposition factions totaling no more than 15,000 men, the United States managed to evict a force of 60,000 Taliban fighters and the regime. It was in essence an air campaign conducted in conjunction with, and supported by, Special Operations Forces and friendly indigenous fighters. It required a relatively limited operation of one hundred combat sorties a day, amounting to 38,000 sorties flown. Outside Afghanistan, a U.S./UK force of approximately 60,000 men, dispersed over 267 bases on thirty locations in fifteen countries, supported this operation. The United States lost thirty men.[110] The use of precision-guided missiles increased again, this time up to 60 percent, indicating that these had become the norm.

A crucial ingredient was the unprecedented integration of ground-air communications.[111] A network of sensors and communication systems glued together combat aircraft, dispersed air bases, command centers, and special forces. In the opening phase of the air campaign, the United States struck fixed targets (roads, bridges, and command facilities), limiting the Taliban's ability to communicate, move, disperse, reconverge, and attack unobserved and unhindered. Afterwards, and because of the insertion into the area of special forces equipped with data links and acting as forward air controllers, attention shifted to so-called emerging targets such as small Taliban troop contingents.[112] Midway in the operation, flex-targeting dominated; 80 percent of sorties took off without specific assigned targets. Instead, Joint Surveillance Target Attack Radar System, unmanned aerial vehicles, and special forces acted as eyes, spotting pop-up targets and relaying time-sensitive, up-to-date, accurate target information to attack platforms inbound or already circling in the vicinity, and handed the aircraft off to the forward air controller. This offered a stunning reaction capability, with response times sometimes down to several minutes, and averaging only twenty minutes.

Operation Iraqi Freedom

Operation Iraqi Freedom reaped the benefits of the experience of Operation Enduring Freedom. The flawed Operation Anaconda of March 2002, where the 10th Mountain Division blatantly neglected to involve the U.S. Air Force in its planning process, had reaffirmed that close cooperation between air and land units was essential. It resulted in improved cooperation between the U.S. Army and U.S. Air Force, greater attention to training and equipping tactical air control teams, and a larger number of such teams. U.S. forces also increased air-land integration at ever-lower tactical levels by assigning air liaison officers and joint tactical air controllers to companies, even to the platoon level if necessary. The conventional phase of Iraqi Freedom sought to capitalize on these developments. While the legitimacy, political and strategic soundness, neoconservative ideological undertones, and neglect of postconflict planning have justifiably drawn heavy criticism, this should not distract from some stunning military achievements.

In the southern portion of the battlespace, the Pentagon relied on precision bombing, a small, fast-moving ground attack force, and the combination of Special Operations Forces and airpower.[113] Admittedly, the command performance of the Iraqi regime was dismal. Still, the rapid advance and subsequent victory of the U.S. forces confirmed that the combination of innovative concepts and power projection with high-technology forces for advanced expeditionary warfare could achieve objectives at the cost of astonishingly low numbers of friendly casualties and modest collateral damage. Operating from a virtual sanctuary, airpower once again proved devastatingly effective against conventional forces. Air-ground surveillance systems, unmanned aircraft, and Special Operations Forces located conventional Iraqi forces while a continuous stream of fighter aircraft delivered ordnance on accurate target locations. Within the Combined Air and Space Operations Center, the Time Sensitive Targeting Cell responded to emerging information on important targets by retasking orbiting fighters or bombers. Often it took approximately twelve minutes to destroy a confirmed target; in some cases attacks occurred a mere five minutes after detection. In the west and north of Iraq large numbers of Special Operations Forces teams operated as part of a closely integrated network with airborne sensors, command nodes, and offensive aircraft to detect and neutralize potential launches of surface-to-surface missiles such as Scuds and to restrict Iraqi freedom of movement on the ground, thus tying down several regular divisions in the north of Iraq.[114]

Networking of forces contributed to the tempo. The combination of intensive air strikes with the highly mobile ground force continued day and night, and small, fast-moving forces defeated larger forces. The United States had only 125,000 forces in Iraq, with only three divisions forming the "spear" of the attack, while Iraqi forces numbered 400,000, including some 100,000 well-trained and well-equipped Republican Guard troops. In a single week, the coalition destroyed a thousand tanks and reduced the Republican Guard by 50 percent. Even urban operations saw enhanced effectiveness of air strikes. Intense intelligence preparations had produced detailed maps featuring codes for individual buildings in specific areas in Baghdad that facilitated close air support coordination with ground troops.

Airpower Explained through *A Discourse*

With all the caveats that such an exercise requires, this brief sketch of recent air operations allows us to explore the connection between Boyd's work and airpower, using the image of the enemy as a complex adaptive system. First of all, merely having airpower adds variety and allows the joint force thus equipped to display new modes of behavior. An army with airpower is a vastly different "beast" than an army without it. Airpower enables forces to develop and exploit an entirely new flank, or sanctuary, over enemy territory to which the opponent must respond. Once air superiority is achieved, it expands freedom of action. Conversely, it can negate certain modes of behavior for the opponent and force the opponent into a purely defensive posture or require the opponent to switch to urban or irregular modes of warfare. It limits the opponent's freedom of action and can oblige it to follow a strategy to which its forces are poorly suited. Simply knowing that it might be observed and subsequently attacked significantly modifies the opponent's behavior, and this holds true for the leaders of a state, commanders of military units, and leaders of guerrilla and terrorist groups.[115] The very presence of airpower thus changes the nature of the fight and the range of options for the warring parties.

Second, airpower plays a large role as a lever to manipulate the informational dimension of war. As demonstrated in the runup to the joint land campaign of Desert Storm, airpower may provide a screen behind which ground troops can safely configure, move, relocate, mass, and subsequently surprise and overwhelm enemy troops. Airpower can prevent an opponent from "reading" a pattern in operations, and thereby increase the level of uncertainty with which the opponent must deal. It can also produce a hugely advantageous information differential in conducting surveillance and reconnaissance missions. As air campaigns in the past two decades have demonstrated, Western airpower can quickly blind an opponent's political and military command capabilities. At the strategic, operational, and tactical levels it can isolate commanders and troops in the information domain, thus magnifying surprise and uncertainty, as Iraqi and Taliban troops and their leadership discovered. Overall, an airpower advantage can have paralyzing effects on the "brain" function of the enemy system.

Airpower can subsequently play a leading role in any of the three categories of conflict—attrition, maneuver, moral—Boyd described, with the aim of rapidly degrading cohesion in the opponent's armed forces. As amply demonstrated during Desert Storm and Operation Iraqi Freedom, once air superiority is achieved, precision airpower can inflict unprecedented levels of debilitating attrition of enemy forces and military capabilities in a short time. This also eliminates specific kinds of threats—airpower, armor, artillery—and once again diminishes the range of options available to the opposing commander. Indeed, although both Desert Storm and Iraqi Freedom contained elements of maneuver warfare, the vast disparity in airpower capabilities allowed the United States to use airpower for attrition-style warfare in lieu of ground operations, enabling ground forces to conduct high-tempo maneuver actions that in turn produced shock and uncertainty among enemy troops and commanders.

Airpower can also demoralize troops and degrade their trust in their leadership, as happened during Desert Storm and Iraqi Freedom, according to interviews with prisoners of war. It can destroy infrastructure and logistics, thereby isolating units physically, which can have an impact on morale. It can also negate maneuvers as well as massing of troops and armor, because constant air surveillance and rapid precision strikes make both fatal options. This reduces the opponent's operational tempo and makes the opponent more predictable, which in turn reduces uncertainty for friendly ground troops.

With air superiority, a blind enemy, constant surveillance, and precision weapons, airpower permits a much higher operational tempo and enhanced intensity at the campaign level, which in turn increases uncertainty for the opponent. It overwhelms the opponent's system and may outpace its capacity to cope with damage to that system. Whether this can produce victory by itself is a moot question; from Boyd's perspective it would suffice if the enemy system is sufficiently downgraded that friendly forces can achieve victory with acceptable risks.

As Operation Enduring Freedom, the subsequent counterinsurgency campaign in Afghanistan, and the Israel Defense Force Operation Cast Lead of 2008 have demonstrated, the capabilities for dynamic and time-sensitive targeting developed over the past decade even enable regular forces to succeed in tactical engagements against small bands of irregular fighters in mountainous or urban terrain, contexts that used to be prohibitively costly. The cognitive networking of sensors and shooters has greatly enhanced adaptability.

Moreover, as the debate on the "Afghan model" has suggested, the combination of small groups of special forces, aided by proxy forces and cognitively linked to a network of surveillance platforms and airborne fire support assets, offers significant combat capability that can produce operational- and even strategic-level effects, as demonstrated during Operation Enduring Freedom, the operations in northern Iraq in 2003, and (in slightly modified form) during Operation Unified Protector in 2011.[116]

Table 2.1 summarizes the effects and functions of airpower in the framework of Boyd's three categories of conflict.

Interestingly, Boyd also recognized that misapplication of force, or the perception of disproportional damage, might erode public support, as it would seem counter to our own moral standards. Over the past two decades all air campaigns have resulted in claims that the West did not abide by its own standards and codes of conduct as laid out in international law and the Law of Armed Conflict. Opponents have shown a keen awareness of Western sensitivities for such claims and have often exploited this moral dimension through cunning manipulation of the media. The debate on drone warfare is only the latest manifestation of this phenomenon. Whereas in 2002 one

TABLE 2.1 Contribution of Airpower to Enemy Organizational Collapse in Three Operations

AIM	METHOD AND MANIFESTATION	DESERT STORM	ENDURING FREEDOM	IRAQI FREEDOM
Exploit destructive force	Weapons (mechanical, chemical, biological, nuclear, etc.) that kill, maim, and/or otherwise generate widespread destruction	✓	✓	✓
Exploit protection	Ability to minimize the concentrated and explosive expression of destructive force by taking cover behind natural or manmade obstacles, by dispersion of people and resources, and by being obscure using camouflage, smoke, etc., together with cover and dispersion	✓	✓	✓
Exploit mobility	Speed or rapidity to focus destructive force or move away from adversary's destructive focus	✓	✓	✓
Produce frightful attrition	Via widespread destruction as a basis for 1) breaking enemy's will to resist and 2) seizing and holding terrain objectives	✓	✓	✓
Exploit ambiguity	Alternative or competing impressions of events	✓	✓	✓
Exploit deception	Erroneous impressions of events	✓	✓	✓
Exploit novelty	Impressions associated with events/ideas that are unfamiliar or have not been experienced before	✓	✓	✓
Exploit fast transient maneuvers	Irregular and rapid/abrupt shift from one maneuver event/state to another	✓	✓	✓
Produce disorientation	Mismatch between events one (seemingly) observes or anticipates and events (or efforts) one must react or adapt to	✓	✓	✓
Produce surprise	Disorientation generated by perceiving extreme change (or events or efforts) over a short period of time	✓	✓	✓
Produce shock	Paralyzing state of disorientation generated by extreme or violent change (or of events or efforts) over a short period of time	✓	✓	✓
Produce disruption	State of being split apart, broken up, or torn asunder	✓	✓	
Exploit menace	Impressions of danger to one's well-being and survival	✓	✓	✓
Exploit uncertainty	Impressions, or atmosphere, generated by events that appear ambiguous, erratic, contradictory, unfamiliar, chaotic, etc.	✓	✓	✓
Create mistrust	Atmosphere of doubt and suspicion that loosens human bonds among members of an organic whole or between organic wholes	✓	✓	✓
Payoff	Surface fear, anxiety, and alienation to generate many non-cooperative centers of gravity, as well as subvert those that adversary depends upon, thereby magnifying internal friction	✓	✓	

scholar labeled the Western way of war with precision weapons as "humane warfare" because of its ability to avoid widespread indiscriminate destruction, another scholar later labeled it as "risk transfer warfare" because in modern campaigns neither Western troops nor Western societies run any risks in "stand-off warfare."[117] Indeed, as Boyd predicted, confronted with Western airpower, dictators such as Milosevic and Saddam Hussein, and groups such as Hezbollah, Hamas, and the Taliban, have tried to open up a new front in the media and thereby exploit what Boyd terms the moral domain.

Analysis

This examination of recent campaigns suggests that Boyd's work can express and explain the functions and effects of airpower, the dynamics it may unleash, and the logic of its contribution to achieving military objectives. Boyd's metaphor of the enemy as a complex adaptive system and of war as a confrontation between two or more such systems, with all the nonlinear effects and emergent behavior that mark such systems, offers a coherent framework. By focusing on adaptability, Boyd also offers new terms and original reframing of familiar dynamics. His work opens vistas on dynamics, logic, and possibilities to bring about organizational degradation and perhaps ultimately collapse and surrender, but not necessarily on what forces *should* do. It is explanatory rather than prescriptive or deterministic, but it does offer plausible generalizations based on both a survey of historical case studies and insights concerning the workings of open systems.

Drawing on Boyd's work and the brief sketch of recent experiences, we can express the function of airpower in terms of its ability to create effects in the physical, mental, and moral domains, in (from a historical perspective) incredibly short time spans, and at all levels, both at the front lines and deep in enemy territory. The physical actions impact the cognitive functions of the opponent at all levels, reduce the variety in the enemy's system, destroy relationships, and thus lower the overall energy level of the system and its ability to cope with the inflicted damage.

While Boyd's work certainly implies that military theorists would focus particularly on the strategic level of an opponent's system and therefore target the political and strategic leadership, his work does not offer a specific argument to that effect. His work does not refer to pure strategic paralysis, but rather to systemic paralysis that might or might not result in the desired strategic outcome. For example, the success of the coercive campaign against Serbia in 1999 did not result from strategic paralysis or an exclusive focus on leadership targets but from the cumulative effects of strikes against a variety of targets over time, combined with changes in the political context (triggered by the continued bombing) that produced sufficient political pressure, as well as exasperation and hopelessness, among the Serbs.[118] Interestingly, system-wide, high-intensity attrition seems a prime advantage that contemporary airpower can offer (and gradual attrition if politically necessitated). As the examples above suggest, in practice such attrition has usually led to success.

Synthesis

Boyd's work constitutes an eclectic search for patterns of winning and losing through a survey of military history; an argument against techno-fetishism and an attritionist, deterministic military mindset; a rediscovery of the mental/moral dimensions of war; a philosophy of command and control; a redefinition of strategy; a search for the essence of strategic interaction; a plea for organizational learning and adaptability; and, finally, an argument on thinking strategically. By giving epistemology, uncertainty, and cognitive processes central stage, as I have argued elsewhere, Boyd became the first postmodern strategist.[119]

The OODA loop, including the simple model, actually presents a rich array of ideas and levers to manipulate, well beyond speed and information superiority. The comprehensive drawing of the OODA loop and a complete reading of Boyd's work suggest strongly that the fundamental theme that connects the various parts of *A Discourse* centers on maintaining the ability to adapt while denying that ability to the opponent. As Boyd noted in *Patterns,* "Adaptability is the power to adjust or change in order to cope with new and unforeseen circumstances," and "in dealing with uncertainty it seems to be the right counterweight."[120]

One must be careful not to read too much into Boyd's work as a theory of, or for, airpower. It includes no arguments for focusing on a particular set of targets. Yet his conceptualization of the opponent (and oneself) not merely as *a* system but as a *complex adaptive system* and of war as a contest between two or more such systems has proven fruitful in suggesting new avenues for engaging with an opponent. His novel expression of the core dynamic in war and his three categories of conflict help explain how airpower has contributed to contemporary joint operations, since they express airpower as a capacity to influence the coherent functioning of the enemy's open system and turn it into a closed system. As Operations Desert Storm, Enduring Freedom, and Iraqi Freedom suggest, airpower has huge potential to induce organizational collapse at the tactical, operational, and strategic levels, and sometimes at the political level. This does not address or diminish the role of either land or maritime forces; it merely suggests that modern airpower provides a significant tool to affect the opponent's functioning as a complex adaptive system. Thus Boyd may not qualify as an airpower theorist, but he certainly has given scholars new and fruitful perspectives on the dynamics of armed conflict, perspectives that also have significant value for understanding airpower in the postmodern era.

CHAPTER THREE

Smart Strategy, Smart Airpower

John A. Warden III

When a new technology appears in war, just as in business, certain advantages in cost or efficiency, albeit initially marginal, may become evident almost immediately. Conversely, decades or even centuries may pass before most people realize that the new technology does not merely substitute for the old, but rather offers the opportunity to move into new, previously unavailable or even unimaginable dimensions. Thus, Thomas Newcomen invented the steam engine to be a replacement for water pumps driven by animals or men; Guglielmo Marconi developed a device to bridge gaps in the telegraph system; the computer initially was primarily designed to improve on human adding and subtracting; and the Internet improved on the post office as a way for government scientists to exchange information with their colleagues. In each of these cases and in thousands more, what began as a substitute for an old way of doing something ended up enabling new concepts of operations never before envisioned. So it is with the airplane and the more encompassing term "airpower."

Airpower saw its first significant application in World War I. As with most new technologies, its initial use focused on augmenting or improving well-established war functions. Thus, airplanes initially executed reconnaissance operations that were difficult for cavalry or ground patrols to conduct. Then, just as enemy cavalry had opposed reconnaissance missions, airplanes took on a counter-reconnaissance role as they tried to shoot down the observer aircraft. This in turn led to a new role centered on protection of the observer aircraft and to battles for control of the air that differed little from skirmishes between horse cavalry except for the dimension in which they were executed. Much as cavalry had harassed opposing infantry, airplanes then began dropping small bombs on enemy troops; as their performance improved, they bombarded targets well behind the lines, the airborne equivalent of deep cavalry raids. As the war continued, bombs fell in greater numbers closer to the fronts to augment traditional artillery. Thus, the roles performed by airpower expanded, but remained analogous to those of its ground-based counterparts.

Before the war, when he was a professor of strategy at the French War College, Marshal Ferdinand Foch reportedly stated that "airplanes are interesting toys but of no military value."[1] As far as is known he did not repeat this comment after the war. Nevertheless, with the exception of a few really advanced thinkers such as the Italian Giulio Douhet and the American Billy Mitchell, the majority of military practitioners continued to see the airplane merely as an enhancement of existing capabilities. Even today, few understand that airpower can and should fundamentally change the very nature of war. Strict adherents of Carl von Clausewitz, of course, maintain that the nature of war can never change, but they are correct only if the "nature of war" simply means the use of force against an opponent.

The first known combat employment of the airplane took place over Libya during the Italo-Turkish war of 1911.[2] In the intervening years, the range, speed, payload, and accuracy of aircraft have improved substantially, and airpower has profoundly influenced the outcome of every conflict following its first major application in World War I. Despite airpower's demonstrated successes, however, most military leaders still see airpower primarily as a means to improve or facilitate old ways of war rather than as the path to revolutionary change of enormous value to those who possess it.

Airpower can never realize its real potential so long as it is bound to an anachronistic view of war described using an anachronistic vocabulary. On the contrary, for airpower truly to come of age, it must do so in the context of a modern concept of war in which the use of force is associated as directly as possible with end-game strategic objectives, not with the act of fighting, which has classically defined the "nature of war." If this is to happen, it will be necessary for the operators of airpower to understand, believe, and teach end-game strategy as the foundation of airpower. Failure to do so will condemn airpower to suboptimization and will deprive its owners of the ability to use force in such a dramatically different way as to achieve national objectives quickly and at minimum cost. To succeed, airpower advocates must stop trying to use airpower as a substitute for its military predecessors (although it can do this quite nicely), must connect airpower directly to strategic end-games, must adopt a new vocabulary to match its promise, and must become serious promoters not of machines, but of ideas.

This chapter first proposes a general strategic approach to warfare. It discusses the key elements of such a strategy and describes methods that commanders and planners can apply in the context of any conflict envisioned. At the conclusion of the general strategy presentation, the discussion returns to airpower, showing how it has a unique capability to deliver the effects necessary to realize this transformative vision of system-level warfare.

The Past Is Not the Future

War seems intrinsic to the human condition, although history only supplies reasonably detailed knowledge about the wars of the last several thousand years. The majority of these conflicts took place between opposing land forces, and thus the bulk of human thinking and writing has focused on the land aspect of conflict, to the point of fixation. By comparison, relatively little has been written about sea power, although it often played a crucial role in the outcome of conflicts dating back at least to the ancient Greeks. As an example of the fixation on land fighting, consider Clausewitz's exhaustive book *On War,* in which the key role of sea power in the defeat of Napoleon is conspicuous by its absence, which suggests that a better title might have been *On Continental Armies in the Era of Black Powder.* Land operations have so dominated the study of war that war itself has come to be defined almost exclusively as the clash of armies. Those clashes, those battles, became not merely the measure of success but events to be sought and desired. As Clausewitz said,

> Combat is the only effective force in war; its aim is to destroy the enemy's forces as a means to a further end. . . . It follows that the destruction of the enemy's forces underlies all military actions; all plans are ultimately based on it, resting on it like an arch on its abutment. . . . The decision by arms is for all major and minor operations in war what cash payment is in commerce. . . . Thus it is evident that destruction of the enemy forces is always the superior, more effective means, with which others cannot compete.[3]

Our purpose is not to dissect Clausewitz's writings. As a writer still much studied and in many ways as the high priest of Western military thought for a century and a half, however, Clausewitz exemplifies how most people, including heads of state and senior military officers, think about war. War, to them, is inevitably the clash of arms; as Clausewitz says in the quote above, "the destruction of the enemy's forces underlies all military actions . . . that destruction of the enemy forces is always the superior, more effective means, with which others cannot compete." It is amazing how this idea has remained so embedded in our thinking and culture, especially in light of a number of historical examples where wars were either won or significantly influenced by some other means. Readers need no reminder that one of the world's truly great empires grew largely on the back of a Royal Navy that frequently won "wars"—or prevented them—sometimes by its mere presence.

Adherents of the Clausewitzian view often overlook that even Clausewitz states that the "aim is to destroy the enemy's forces as a means to a further end." Despite this recognition, he and his followers then focused their thinking, writing, and war plans on fighting. Unfortunately, most military leaders still only pay lip service to "the further end" while remaining fixated on battle.

If we really want to rethink war and escape from the shackles of a bygone era, the vocabulary of war needs to change dramatically. Among other things, words and phrases from the past such as "fighting," "battle," "shape the battlefield," "battlespace," and "warfighter" should be expunged so as to elevate the "end" to the forefront of consideration. In other words, it is time to bury thousands of years of bloody war stories, heroic as they were, and start looking at war—and eventually airpower—from its end point, which by definition means from a strategic perspective.

The True Essence of War: Strategy

The very word "strategy" has multiple definitions. In this chapter, it means the totality of how an organization such as a nation or any other group intends to move from one condition to another.[4] Although the direction of movement could be positive or negative, our assumption here will be that it is positive. That is, the group intends over some period of time to improve its lot. For a nation, that improvement could range from world conquest to survival in the face of attack.

Strategy includes objectives (conquest, survival) with concomitant acceptable cost and risk, identification of what needs to change internally and externally to achieve the objectives (but not the tactics to make it happen) against which resources will be applied, control of time, and transition from one phase to another. Sometimes, people will use the term as though it applied to only one of these components; thus, "their strategy was world conquest," "they employed a strategy of attrition," "their strategy was to trade space for time," or "their strategy was to die fighting." All of these might be part of a strategy, but used individually they are of little value.

The concept of strategic unity is most important in thinking about the use of force, as it is common for the military planner to begin by applying resources against one part of an opponent (normally the opponent's military or equivalent) with little concern or little idea of the other strategy parameters. Likewise, it is common for political leaders to make pronouncements about objectives ("make the world safe for democracy") with no serious consideration for the other necessary elements of strategy. The result of this disjointed approach to war is all too frequently a rapid descent into suboptimized, locally generated tactical actions that are unlikely to move the organization to its desired state at an acceptable cost and risk.

To create true strategic unity, it is necessary for the leadership of the state to accept responsibility for crafting a complete strategy that entails a clear description of objectives, exactly how much risk it is willing to accept and what it is willing to pay in blood and treasure, what aspects of the opponents and self need to change, how much time is available to achieve success, and how the contest will be brought to a close. Only by going through such a complete process is it possible to determine the relative probability of success. For example, if the objective is world conquest but there is no careful thought as to what opposition would need to be overcome, no careful calculation of how much probable operations will cost or how much is available to spend, and no timeline or clear description of what happens after successful (or unsuccessful) world conquest, the chances of developing and executing a successful plan are remote.

Except in rare circumstances, executing a strategic plan will require effort by multiple parts of the organization (military, diplomatic, political, financial, economic). Assignment of specific strategic responsibilities (*not execution tactics*) is the job of the political leadership and cannot be delegated, although it can and should be jointly developed. In fact, only through joint development (open planning with wide participation) is a good strategy likely to materialize and be well enough understood by enough parts of the group to make successful execution likely. It is also in planning sessions of this kind where it should become clear whether military force has applicability. If a state wants to be successful in any strategic endeavor, and most especially war, it must genuinely involve in the high-level planning those who will execute military and other operations.

Strategy as described above appears quite complex, but for the present purpose we can simplify it considerably. At the most basic level, building a strategy consists of formulating answers to four strategic questions: Where? What? How? Exit? The answers taken together, not any single answer, constitute real, comprehensive, unified strategy.

Where do we want to be at some specific point in the (postwar) future and what are we willing to pay? For simplicity, we call this a future picture. It constitutes our strategic objectives along with the risk and cost acceptable to realize the objectives.

What are we going to put our resources against to create the conditions that enable us to realize the future picture? At the highest level of analysis, this process begins with identifying the systems (enemy and friendly) that must change. The next level of analysis determines the centers of gravity (the control points or the leverage points) against which to apply tangible resources to force the desired system change.

How, and in what time frame, must we affect the entities against which we will apply our resources? In this step, commanders will eventually make decisions about tactics, but they must start by determining the applicable time frame and the order of attacks against centers of gravity within that time frame. Strategists should delay the choice of tactics—a bomb, a bullet, a bribe, a torpedo, a dollar, a gift—until they understand every other factor.

Exit. In what manner will we move on following success or failure? The ultimate goal of any war is "winning the peace," creating a sustainable postwar world. Even when combatants achieve their aims, moving on from success is difficult, and the plans for a transition to peace must be analyzed at least as carefully as the future picture and the decision to go to war. Even more dangerous is failure to plan for the high likelihood that individual actions throughout a campaign may fail.

The Future Picture (Where?)

Some states and organizations go to war deliberately for a variety of purposes; some wars are accidental, and roughly half are reactions to an attack. Regardless of the reason, success or failure and accompanying rewards or penalties will happen after the war starts, in other words, in the future. Since reward or penalty happens in the future, planning itself should start from the future and work back to the present.

In fact, most thinking about war operations fails to start from the future. In many cases, it actually starts from the past (revanchism, or the recovery of long-lost territory or pride). At best it begins sequentially from the prevailing initial conditions (the present), focuses on the opposing military forces without much regard or understanding of the overall war objectives, and tends to be philosophically reactive instead of proactive. For example, a provocation arises and the immediate impulse is to counter it where it happened. Or in an offensive war situation, the thinking typically starts with the nearest opposing military force. Although these approaches may be easy and may bring emotional satisfaction, they are rarely strategically useful, as they simply set in motion a train of events the conclusion of which is unclear. As an example, immediately after Iraq's invasion of Kuwait in 1990, the U.S. response most frequently proposed was to attack the Iraqi army directly in Kuwait, with no real thought given to any outcome beyond, perhaps, expulsion of the Iraqis from Kuwait. Similarly, when Serbia began driving the ethnic Albanians out of Kosovo, the North Atlantic Treaty Organization's (NATO's) initial reaction was to attack the Serbian soldiers and tanks involved, again without considering the efficacy or the longer-term consequences of such an approach. An example of this kind of thinking in offensive wars took place in World War II when the Germans started their major operations with an attack on Poland. Although the attack succeeded from a military standpoint, there were no prior decisions for follow-on operations against the other opponents that resulted from the Polish operation.

Failure to plan from the future is like trying to drive down the highway by looking in the rearview mirror; the focus is on what was. It almost always leads to tactical solutions devoid of strategic context and serious thinking about the future. It also favors the use of the military forces immediately available or those that have represented the traditional solutions to a particular issue. Yet the true goal of any armed conflict—winning the peace—intrinsically lies in the future; thus planning from the future is imperative.

Failure to plan from the future almost always leads to bottom-up (sequential) planning. It consists of thinking serially about issues, beginning with the closest one in time or space. For example, in planning the 1914 invasion of France, the Germans roughly followed the 1905 Schlieffen Plan, designed to defeat the French army in a battle of annihilation within about six weeks of initiating hostilities. The starting premise (the "bottom" condition from which planning began) was the French army and the Clausewitzian idea that a big battle and defeat of the French army would lead automatically to war success (which was largely undefined beyond a repetition of the 1870 French surrender at Sedan). The Germans intended to encircle the French by launching a strong attack from their right wing through the Ardennes while the much weaker left wing held in the formerly French province of Alsace-Lorraine. The plan was conceptually sound from a limited military perspective, but it had a fatal flaw that resulted from its bottom-up approach: it ignored the possibility that the British would honor their obligations under the London Treaty of 1839 to aid

Belgium if its neutrality was violated. The intervention of the British Expeditionary Force slowed the right wing's advance sufficiently to allow the French to reinforce.

More recently, when the United States attacked Iraq in 2003, the plan focused strictly on taking Baghdad. In this case, the starting premise (the "bottom" condition) was an assumption that the occupation of Baghdad would lead to Iraqi acquiescence, that the primary force would be the U.S. Army that would defeat the Iraqi army en route to Baghdad, and that the Army would proceed from existing staging areas and bases in Kuwait. While U.S. forces succeeded in defeating the Iraqi army and in capturing the city, no plans existed for the next stage of what became a very long and expensive war. The war was planned "bottom-up" to a Clausewitzian end point of defeat of the Iraqi army.

The future picture captures the planner's vision of the desirable future without reference to the current situation or to the possible difficulties of bringing that future about. It functions as a powerful, continuous beacon that informs every decision and every act. It describes an end state; it does not prescribe actions. As planners develop a future picture, they cannot and need not know what actions might be necessary to realize it. In fact, any detour into thinking about tactics would almost certainly steer thinking into suboptimal channels. Eventually—but much later in the process—planners must tie the target for every bomb, missile, or other weapon to the future picture.

The first step in building an effective strategy involves constructing future pictures set in a time frame well after the end of the envisioned war. Strategists must create two future pictures: one for their own state or group and the other for the opponent.[5] Many might find it surprising that effective war planning first examines the future of their own side, but the reason is simple: their own future is paramount, not that of the enemy. Once strategists have clearly envisioned and charted their own future, they can create a future for the enemy that is compatible with theirs.

For the future picture to be useful and executable, it must include clear measures for both sides that indicate progress toward achieving it. These measures must be strategic, not tactical; no sane leaders would want to win battles only to find that the conflict had destroyed or seriously weakened their own country in the process.[6] King Pyrrhus of Epirus learned this lesson too late after paying an impossible price in casualties while winning multiple battles during his five years of war with Rome in the early third century BC. Pyrrhus focused his planning and fighting on defeating Roman (and other) forces in a number of battles. He actually succeeded brilliantly at a battle level, but in the process so drained the resources of his own country that it became an inviting target for the Romans after Pyrrhus died at Argos. So dramatic was the lesson that two thousand years later "Pyrrhic victory" remains part of our language. Pyrrhus lacked a unifying future picture that would have allowed him to choose battles, risks, and costs with requisite care. Two thousand years later, in World War I, the British, French, and Germans emulated Pyrrhus by sacrificing huge numbers of men in horrific battles. Although these battles led to occasional great tactical successes as measured by captured territory or exchange ratios, the battle winners were unable or unwilling to deal with the obvious strategic damage they were doing to themselves and with the corresponding lack of strategic measures indicating any real progress against their opponents.

True leaders bear primary responsibility for the aftermath of military operations, not only for the outcomes of the operations themselves, but also for helping to create a viable postwar world. Commanders may think they have no alternatives to battle, but battles are the means to an end, not the ends in themselves that they almost inevitably become. Further, it is possible to win the peace and lose all the battles, as North Vietnam did. In addition, it is possible (and certainly desirable) to win a war without fighting a single battle and with as little damage to the opponent as is necessary. As the great Marshal of France, Maurice de Saxe, said in 1748, "I am even convinced that a skillful general can wage war his whole life without being forced into one [pitched battle]."[7] Similarly, Sun Tzu said, "Supreme excellence consists in breaking the enemy's resistance without fighting."[8] Both these statements represent the antithesis of Clausewitz and ought to be the first option considered by the good commander. Indeed, the concept of winning wars without fighting was one of the cornerstones of the foreign policy of the Eastern Roman Empire, the longest lasting empire in human history.[9]

Envisioning the geopolitical future picture begins with a precise understanding of what the strategist wishes to accomplish through military or other operations. In war, this is (or at least should be) very simple: win the peace. This concept is of extreme importance when crafting grand strategy, especially with a view toward airpower application. It is easy to name many famous generals, but most of them earned their fame by winning battles. For example, military commanders still study and copy Hannibal's brilliant victory at Cannae, where he annihilated a large Roman army in a few hours thinking his victory would make him master of the world. He then refused the advice of Maharbal to advance on Rome, the true center of gravity of his mortal foe. Thus, his battle victory availed him little. Within a few years Rome had recovered and managed to expel Hannibal from the Italian Peninsula. Hannibal then lost a great battle outside the walls of Carthage to a Roman army commanded by Scipio Africanus, one of the few Roman survivors of Cannae, and suffered exile not long thereafter. His reliance on battle set into motion a chain of events that inexorably led decades later to a vengeful Rome deciding to obliterate Carthage and literally destroy an entire civilization. In sum, although he won a great battle, Hannibal failed to win the peace and eventually his country perished. By that criterion, Hannibal—and many centuries later Napoleon—could be considered among the worst generals of all time, despite their felicitous ability to deal with tactical issues.

The strategy process starts with creation of a future picture, the long-term objective. In essence, a future picture is a high-resolution, measurable, and compelling portrayal of both sides at a point in time after the war ends. Calling it a future picture emphasizes its reality and underlines that it is something concrete, not a wish nor a dream. A well-crafted future picture does not change except under the most unusual circumstances. An excellent example of the perils of violating this concept comes from the U.S. experience in the Korean War. Prior to June 1950, when the North Koreans made a surprise attack on South Korea, the United States had declared that Korea was outside its sphere of interest. Immediately following the attack, the United States suddenly decided that it had vital interests in Korea, and its future picture then became one of avoiding defeat and expulsion from the peninsula. After Douglas MacArthur's brilliant amphibious landing far behind North Korean lines and the U.S. sweep to the Yalu River on the Chinese border, Washington changed its future picture to one of a unified Korea. But when the Chinese drove the U.S. and United Nations forces back down the peninsula, yet another future picture emerged for the United States: survival! Finally, after stabilizing the line on the 38th parallel, Washington established yet a fourth future picture: a Korea divided by the 38th parallel, exactly the situation when the war began.

Failure to have a constant future picture from the beginning meant that the United States would always be reacting to the tactical situation and would not undertake key strategic initiatives. For instance the United States, if the second future picture had been in force from the beginning, might have worked to forestall Chinese intervention. Any of the four future pictures was arguably acceptable and feasible, but constant changes made serious planning and execution virtually impossible. Thus, changes in the future picture at high strategic levels virtually guaranteed dissension, confusion, and failure.

Strategists must remain focused on the future picture and must concentrate not on winning the war, but on achieving a better state of peace. This means looking beyond today's battles and targets to tomorrow's rewards. It means carefully associating targets with the peace that will follow the war; in essence, it means not confusing ends and means.

Future pictures must also have time frames for realization. Without an associated time element, even excellent goals or objectives become mere words, devoid of strategic content, because the absence of a time frame makes it impossible to prioritize resources or to measure progress. At a minimum, the time frame must extend beyond the end of the planned war. Failure to establish a long enough time frame limits strategic options and leads to actions that may appear productive in the short term but later create difficult or insoluble problems. As previously noted, German violation of Belgian neutrality in World War I certainly expedited movement of the German right wing and achieved its goals in the short term, but it also brought Britain into the war, which proved disastrous for Germany.

War-associated future pictures must have high resolution because real forces must realize them in the real world, and realization of a low-resolution future picture is nearly impossible. For example, President Wilson's request to the Senate for a declaration of war against Germany in 1917 really had but one component: "to make the world safe for democracy." Such an abstract goal, no matter how laudable, simply lacked the resolution to guide General Pershing, Secretary of State Bryan, and dozens, if not hundreds, of decision makers in the government and the military. The number of specific features of the desired peace necessary to achieve the appropriate degree of resolution will depend on the circumstances, but it will always be more than one and could easily be as many as a dozen.

Elements of a Future Picture

As noted, planners must envision a future picture for their own state or group as well as for the opponent, and they will find it far easier to develop and use them as two distinct entities. Like the future picture for the commander's own country, the future picture for an opponent focuses on creating and maintaining a desirable postwar peace. Depending on the perspective, a desirable peace might be one in which the enemy has been entirely obliterated, or one in which the enemy will become a future ally. At one end of this spectrum was Rome's putative future picture for Carthage in the Third Punic War, as expressed so simply by Cato the Censor: *Carthago delenda est—Carthage must be destroyed!* In 146 BC, Rome realized its future picture for Carthage by destroying the city, killing all the men, sending the women and children into slavery, and sowing the fields with salt. Two millennia later, Henry Morgenthau, secretary of the treasury in the Franklin D. Roosevelt administration, proposed and partially implemented a near-Carthaginian future picture of turning a defeated Germany into a pastoral society. Near the other end of the spectrum was Bismarck's intention to ensure that Austria would be able and willing to ally itself with Prussia following the Austro-Prussian War. To realize this future picture, he would not allow Prussian chief of staff Moltke the Elder to annihilate the retreating Austrian army. More recently, a major U.S. objective in the first Gulf War was regional stability, meaning that Iraq could not be a strategic threat to its neighbors. That, in turn, meant weakening but not destroying Iraq as a system, thus allowing the state to remain functional and capable of defending itself but not capable of undertaking new foreign adventures. Leaders must continually test a future picture by asking, and then asking again, if that picture would benefit their own country. This requires them to look carefully at all elements of the picture to ascertain what negative or unexpected consequences they might contain.

Assuming we can create a future picture for ourselves and our opponents, the next two strategy questions of direct interest are the second (What?) and the third (How?). It would seem clear that if we want anything (a future picture) that is different from that which currently exists, the enemy as a system must change to make it happen.

System War and Centers of Gravity (What?)

The objective of war at any level is to make the enemy's behavior compatible with the planner's objectives at an acceptable cost. In the geopolitical world, the opponent—whether a nation-state, a group such as Al-Qaeda, or a tribe—must change in some way to match the planner's future picture (strategic objective). Because the opponent will probably not want to change, something must force it to do so.

First, the enemy must have energy to resist the changes that friendly forces might want to impose. At the highest level of system thinking, energy (for offense or defense) is a function of only two dimensions: the physical and the psychological (or "moral" in older parlance). The physical aspect consists of tangible entities such as people, buildings, communications systems, and weapons. The psychological aspect consists of intangibles such as will, morale, and attitudes. War against a system does not concern itself primarily with the psychology of individuals (although that can be quite important), but rather with the psychology that characterizes the system as a whole. The following equation captures the concept:

$$Energy_{Enemy} = f\,(physical) \times f\,(psychological)$$

This equation is enormously useful for thinking about war operations (in fact, any operations). It shows that if either the physical or the psychological energy is zero, the enemy is paralyzed and unable to attack or defend. The most powerful entity in the world cannot succeed in war if it has no will to attack or defend; conversely, the most determined, most aggressive entity cannot succeed without physical capability.[10]

Against this background, the planner can examine the physical and the psychological components of enemy energy at a high level. In theory, the physical dimension is entirely knowable. With good intelligence, the planner could be aware of almost every physical asset in an enemy entity that contributes to its capability as a system, and those elements generally change little over short time frames (hours, days, or weeks). By contrast, planners can know only a little about the psychological side of the equation, and that side can change dramatically in very short time frames (seconds to minutes). To confirm this observation, consider how rapidly one group of seemingly tranquil people can become a stampeding mob and yet an apparently identical group of people, presented with the same stimuli, may respond calmly and rationally.

Operational planners must understand the indeterminateness of the psychological side of the equation, because relying on changing this side of the equation means relying on the unknowable and the unpredictable. War theories such as coercion and deterrence rest on shaky ground from the start; both depend on the enemy's deciding to act or not to act on the basis of concern over the consequences and costs experienced or anticipated. The study of crowds (politicians, investors, speculators, mobs), however, reveals that today's concerns and fears may become tomorrow's motivators. For example, aerial bombardment of Germany during World War II had a negative impact on morale,[11] although it did not drive Germany to sue for peace. By contrast, Iran's decision to agree to a truce with Iraq in 1988 flowed in part from the decline in system morale induced by Iraq's strategic air and rocket attacks. Strangely, however, the lower morale that contributed to Iran's decision to accept a cease-fire apparently had little impact on support for the clerical leaders of the country. Prior to World War II, many observers had believed that a population would collapse under sustained bombardment. Conversely, few observers of the Iran-Iraq War thought that Iraq's low-level bombardment of Iran would have much of an impact on Iran. In other words, predicting psychological effects in any given situation verges on the impossible.

Coercing people means forcing them to agree to an action by harming them or threatening to harm them. Some people, some nations, and some military units may make dramatic concessions at the first hint of a threat, while others prefer to suffer annihilation rather than concede anything. To compound the problem, the people or nation or unit that today accedes to the slightest threat may tomorrow become intransigent, and vice versa. Thus coercion as a war theory does not stand the test of common sense. Planners must accept that they cannot predict what will happen and therefore should never make deterrence, coercion, decapitation alone, psychological operations alone, or other similar mind-based concepts the heart of their operations.

If both factors in the enemy energy equation were equally unknowable and variable, success in war would be a matter of pure and random chance. Fortunately, planners can reduce risk levels and make reasonable predictions about war outcomes if they focus on the physical aspect of the enemy energy equation and on understanding the enemy as a system. This makes it possible to build operations that create the highest possible probability of success with the resources the attacker is willing and able to commit. Simply, the plan seeks to change the enemy's physical system to match the planner's desires. It is useful to note at this point that planners can apply the concept of system change to an enemy state, a terrorist organization such as Al-Qaeda, an enemy army or air force, or an enemy unit such as a corps or a wing. If an enemy leader decides to negotiate before his system no longer functions adequately, so much the better. The rule is, however, to plan on predictable system change and to treat positive system psychological outcomes as a welcome, but unpredictable, bonus. Reversing the process—depending on psychological outcomes through coercion or deterrence—puts the entire plan in great peril. Once operations designed to produce physical change have begun, however, it makes perfect sense to operate against the psychological side as well, provided the necessary resources are available.

In the original planning for the Gulf War air campaign, the Checkmate group tried to focus on the physical as the primary method of achieving U.S. objectives. In our first presentation to General Schwarzkopf on August 10, 1990, however, I included a briefing slide that stated that our proposed strategic psychological operations were as important as the bombing operations.[12] I would not make this assertion today, although our thinking at the time was simple: while a change of regime in Iraq was not necessary for victory, it would have benefited our postwar position significantly. Because of the unpredictability of system psychology, though, we recognized that the Iraqis might remain loyal to Saddam Hussein regardless of what happened to their country, so clearly we could not count on this aspect for success. Although the purpose of the proposed strategic psychological operations was to induce elements within Iraq to overthrow Saddam, doing so was not necessary to achieve the basic objectives of the war.[13] General Schwarzkopf agreed, but he asked about the impact of not deposing Saddam (or seeing him overthrown). I answered that it would be unfortunate for both the United States and Iraq but would not make too much difference overall; what we planned to do to Iraq as a system would mean it would be at least a decade before Iraq could again pose a strategic threat to its neighbors. General Schwarzkopf replied that if the United States could get a decade of return for what we believed would be a very low-cost war, he would be delighted.

System warfare against the physical side of an opponent provides the most positive resolution of conflicts. To execute it well, however, planners must reverse their normal method of thinking. They must think from the big to the small, from the top down.

Systems

To think appropriately about war (whether conventional or unconventional), planners must think in terms of systems. Everything in a nation or entity is part of a system, and every action by that entity takes place within that system. Therefore, affecting one portion of a system will have some impact on other portions of that system. Usually, however, commanders do not wish to change only one part of a system; they want to bring about a major change in an entire system to realize their future picture for the system. This means that efforts should always focus on systems and particularly on the component entities of the system whose alteration or destruction will create the fastest, most long-lasting, most economical, most satisfactory system change.

Systems have certain common characteristics, such as disparate but interacting elements, information flow across the system to and from its elements, at least a minimum amount of energy, inertia and resistance to change, centers of gravity, and similar patterns of organization. Several of these characteristics merit more detailed examination.

Systems—whether nations, companies, universities, or families—resist change and exhibit the hysteresis effect, a term from mechanics that describes how a material under a deforming force tends to return to approximately its original state when the deforming force stops. Thus, when an action perturbs any system—whether by dropping a bomb on a country, inserting a new product into a market, or introducing a new theory to a university department—the system will respond by opposing the "perturbation" and trying to stop it or negate its effect. Planners should never be surprised when systems act this way; indeed, they should be amazed if a system did not. Experts in all fields have had the experience of trying to teach a new method of accomplishing a task to a group of people. After some time, they believe that the students have learned the new procedure and that they can put their efforts elsewhere. Much to their dismay, however, when they return to the group a few weeks later, they find that the group has reverted to its old ways. It will always do this unless its elastic limits are exceeded. In seeking to influence systems, the objective is normally to exceed their elastic limits and thus compel the systems to stay in the desired new state without further expenditure of effort by the party seeking change.

Centers of Gravity

After determining what the enemy system must become in order to realize the war objectives, planners must identify the targets (What?) that need to be affected to produce the desired system change and lead as directly as possible to the future picture for the opponent. When commanders first look at an enemy, whether a large entity such as a nation or a more dispersed entity such as a terrorist group, the number of potential targets may seem overwhelming and well beyond the reach of available resources. Intuitively, however, they realize that out of hundreds of thousands of possible targets, some small number will be far more important than the rest. These important targets constitute the strategic centers of gravity, also known as leverage points or control points; their health and their actions have a disproportionate impact on the rest of the system.

It is also smart to think about centers of gravity in terms of the number of links they have. The mathematicians Barabási and Bonabeau recently derived the relationship between nodes and the number of links connecting the nodes in a system such as the Internet, revealing a small number of nodes that have many links and a large number of nodes that only have one or two.[14] Clearly, attempts to affect a system such as the Internet would achieve far more if the nodes with many links are targeted rather than the ones with only one or two links. When anything positive or negative happens to a node with multiple links, the effect to some degree spreads to all the nodes to which the target node is linked. Conversely, a change to a node with just one link has relatively little effect on the system as a whole. It is worth noting that even in a large system such as the United States or China, the number of nodes (targets) with multiple links and associated with strategic centers of gravity is small, probably considerably less than a thousand.

Methodology

It makes little sense to expend scarce resources against anything other than centers of gravity, yet the majority of planners in both the military and the commercial world devote practically no time to identifying those centers. Instead they rush to action, believing that if they just act forcefully against the military elements of their opponent they will be victorious. Worse yet, military planners are likely only to think about attacking their military counterparts. Leaders who are lucky or who have infinite resources may indeed achieve some positive outcome in this way, but those who are neither exceptionally lucky nor infinitely wealthy should focus on identifying centers of gravity.

Knowing that centers of gravity exist represents the first step toward effective and efficient operations, but planners need a methodology to find the true centers of gravity. The Five Rings Model (figure 3.1) has proven an exceptionally useful construct for identifying such centers, whether in war, politics, education, or business. The model derives from the macro steps that a state (or any other entity) must take in order to launch an attack on an opponent. For example, if a state wanted to attack another state, the following activities would probably occur: (1) One or more individuals (*leaders* with or without portfolio) would advocate the idea of an attack, find other leaders to help them, or suppress those who opposed the idea. (2) The leaders would develop or put into motion the *processes* necessary to garner support from more organs of the state and to acquire resources, such as arms and ammunition, for the attack. They would set other processes into motion to recruit, train, and equip the forces needed for the attack and would nurture the processes necessary for survival of the state such as communications, food production and distribution, finance, and manufacturing. (3) They would ensure that the nation's roads and other *infrastructure* were adequate for survival and for supporting attack operations. (4) They would take steps either to ensure adequate support from the *population* or to suppress opposition. (5) Finally, the state would send some of its *fielded forces* (almost always a

FIGURE 3.1 THE FIVE RINGS MODEL

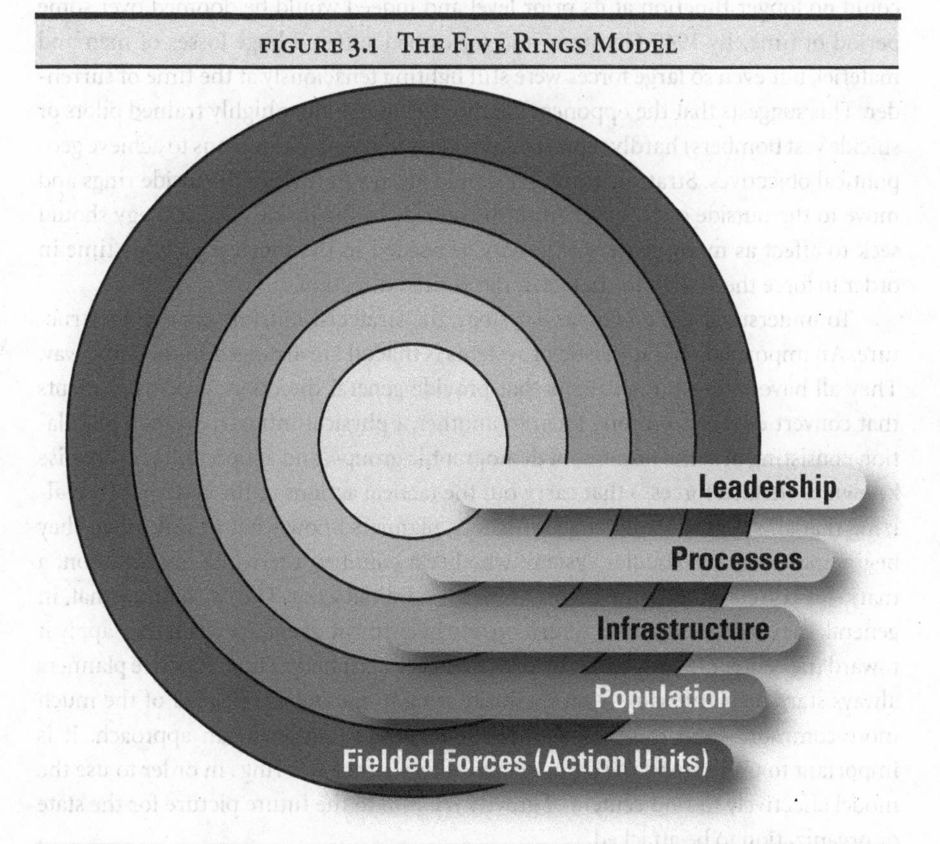

relatively small part of the population at least since the days of the Mongols) to execute the assigned attack. Note that deploying forces is the last step in this simplified process and that the state probably could send additional forces if the initial troops encounter difficulties.

Recognizing this, a strategist can convert these steps into a universal template of a system, any system, from a large nation to a small terrorist group. This template becomes five concentric rings numbered from the inside to the outside and labeled as Ring 1: Leadership; Ring 2: Processes; Ring 3: Infrastructure; Ring 4: Population; and on the outside, Ring 5: Fielded Forces (Action Units). Working backwards through the sequence just described—or, put in a different way, from the outside rings to the inside—would reveal that the state could probably lose the entire force that it had dispatched (which would be part of the fifth ring), but (other things being equal) remain an entirely capable organization. Recall, for example, how quickly the British recovered after military losses in 1781 at Yorktown and in 1941 at Singapore.

Conversely, if the state's leaders had been deposed, or had disappeared, died, or simply lost interest in conflict, or if communication were limited, food production and distribution broken, and movement difficult or impossible, the state (or group) could no longer function at its prior level and indeed would be doomed over some period of time. By 1945 Germany and Japan had suffered huge losses of men and materiel, but even so large forces were still fighting tenaciously at the time of surrender. This suggests that the opponent's armed forces (whether highly trained pilots or suicide vest bombers) hardly represent a viable starting point for plans to achieve geopolitical objectives. Strategic thinking should always start from the inside rings and move to the outside ones, never from the outside to the inside. The strategy should seek to affect as many centers of gravity as needed in the shortest possible time in order to force the system to change in the desired direction.

To understand the enemy as a system, the strategist must determine its structure. An important characteristic of systems is that all are arranged in the same way. They all have leadership elements that provide general direction, process elements that convert energy from one form to another, a physical infrastructure, a population consisting of some number of demographic groups, and action units (otherwise known as "fielded forces") that carry out the tactical actions of the system.[15] By realizing that all systems are arranged this way, planners know what to seek when they begin analyzing a particular system, whether a country, a terrorist organization, a market, a company, an army corps, or even a criminal gang. They also know that, in general, they will get a greater return on any investment of energy when they apply it toward the center of the system rather than on the periphery. Thus, effective planners always start their thinking from the inside ring to the outside instead of the much more common—and misguided—outside-to-inside Clausewitzian approach. It is important to understand the components of each of the five rings in order to use the model effectively to find centers of gravity *relevant* to the future picture for the state or organization to be attacked.

Ring 1: Leadership. The leadership ring consists of those elements of a system that try to move it in a particular direction. Almost all systems have several leadership elements that rarely have the same motivations, are relatively autonomous, may not have formal titles, may be individuals or entities, and usually provide very high leverage. Who (or what) the leadership ring includes depends on the level of the system being analyzed. For example, at the national level the leadership category would include heads of state, prime ministers, influential cabinet ministers, senior military officers (if they are independently influential at a national level), the key media, the legislative body, nationally influential financiers, well-known clerics (in some countries but not in others), important opposition leaders, and perhaps some think tanks. For a military unit such as a division the leadership would consist of the commander, informal leaders (those who act as leaders but who do not have formal rank or position), and probably the staff.

Ring 2: Processes. The processes ring contains those elements of a system that convert energy from one form to another. At a national level this would probably include electricity, petroleum, communications, finance, transportation, agriculture, etc. In a military division, the key processes might be communications, logistics, and transportation; for a terrorist group they could be communications, finance, training, recruiting, and transportation.[16] The processes ring offers great leverage for system change because a change in this ring will affect much of the rest of the system.

Ring 3: Infrastructure. The infrastructure ring comprises those elements of a system that are relatively stationary and constant. At a national level, they might include roads, bridges, rivers, ports, and airfields. This ring can also include such intangible elements as laws and regulations that have a channeling effect similar to physical infrastructure.

Ring 4: Population. For the purposes of military planning, the population ring can be thought of as the demographic groups to which the people in the system belong. Demographic groups tend to respond to similar stimuli (publications, messages, rewards). In the population ring, the interest is in groups, not individuals. For example, if a revolt by the enemy military was desired, messages (stimuli) that might motivate officers in general to change sides would go out through appropriate means.

Ring 5: Fielded Forces (Action Units). This ring contains those elements in a system that perform tactical tasks. Fielded forces (action units) are the instruments of the system: they have latitude in how to perform a task, but not whether to do it; they execute policy but do not have the authority to create it. Examples of fielded forces include a fighter squadron, an army corps, a flotilla, or (in the business world) a sales force or a manufacturing division. Fielded forces are important, but they are appendages of the state or organization, are resistant to attack, can normally be reconstituted quickly by an intact state system, and are means to an end, not ends in themselves in either attack or defense. They are not the starting point for thinking about war.

Although the greatest impact accrues toward the center of the system, this does not mean that plans can focus only on the innermost ring and seek to "decapitate" the system, a notion that some have derived from the system concept. In certain rare instances such decapitation might achieve the planner's goals, but a central concept of system warfare is to avoid creating single points of failure. If an operation attempts to remove the leader and fails, the enemy will have recognized the strategy and made adjustments to counter it, forcing the attacker to try a different tactic against a system that is now prepared. At the same time, the attacker has moved further into the very dangerous serial world, discussed later.

The preceding discussion should demonstrate why even successful attacks on enemy military forces are unlikely to produce the system change necessary to accomplish the strategy's objectives. If we think about the energy employed to attack centers of gravity in each ring as an "investment," it becomes clear that our return on investment will be relatively high for the first two rings but will fall off dramatically toward the fifth ring.

Once the strategist understands the five-ring structure of systems, it is easy to find the centers of gravity in any system. After reviewing the future pictures for their own country and the opponent and the desired system effects for both, planners should start with the leadership ring and identify elements that will have a disproportionate impact and will advance the realization of the future pictures. They then follow the same process for the remaining four rings.

Planners must not confuse vulnerabilities and centers of gravity and should never begin a system analysis by looking for vulnerabilities. A search for vulnerabilities at a strategic (or operational) level distracts planners from finding the real centers of gravity in the system, whether or not they appear vulnerable to attack. Once planners recognize the elements within the opponent's system that must function in order for the opponent to resist the future picture, the planners can develop attack methodologies specifically geared to affecting their centers of gravity. Further, all actions must aim at influencing the enemy system as a whole. Thus strategists should not plan an attack against industry or infrastructure primarily because of the effect it might or might not have on fielded forces, but for its direct effects on the entire enemy system that will lead as directly as possible to realization of strategic objectives.[17] In a few cases, one or two centers of gravity will suffice, but in most instances the actions should be chosen to affect several centers in a relatively compressed period of time.

To address the opponent's fielded forces at all—which may or may not be necessary—planners can and should use exactly the same methodology they used at the system level. After identifying the objective—which could range from destruction through immobilization to recruitment—they analyze the fielded force as a system and find the relevant centers of gravity, starting with the leadership. Similar to the national level, the number of centers of gravity that the plans must address in this case normally translates into far fewer targets than in the traditional war of attrition against the opponent's men and equipment. Even a large fielded force again presents surprisingly few important targets, probably in the low thousands at most, as in the Iraqi army in Kuwait in 1991.

After identifying the centers of gravity, the planner must decide whether to destroy, build, isolate, convert, or paralyze each center to achieve the desired effect on the opponent, allies, and friendly forces, and how to measure success. Note that destruction or damage is just one of many possible desired effects for centers of gravity. In other words, desired effects may be negative or positive. Regardless of the specific desired effect, it is important that it and the center of gravity to which it applies be as directly related as possible to strategic objectives as opposed to being means to an end such as a fight with opposing fielded forces. Using examples of desired effects by ring makes it easier to understand the details of this step.

Leaders (ring 1): If a strong leader such as an Attila, Napoleon, Bismarck, Hitler, or Bin Laden is taking the opponent in a particular direction, the removal of that leader (and perhaps his close associates) normally results either in a reverse of direction or a significant deceleration. Removal or conversion of a leader (through force, persuasion, or even bribery) would constitute a direct strategic action, as the change in the center of gravity is directly associated with a strategic objective.

Processes (ring 2): A good plan seeks to make it impossible for the enemy to pursue objectives contrary to the desired future picture. For example, in World War I, the Allies imposed a blockade on Germany that Liddell Hart believed was "fundamental" to the outcome of the war. With its supply of food essentially cut off by late 1918, Germany was hard-pressed to survive as a state even though it could still fight at the front; thus the blockade produced a direct strategic effect. Further, as the blockade continued into 1919, it drove the postwar German government to accept the harsh terms of the Treaty of Versailles.[18]

Infrastructure (ring 3): A nation-state or a group needs a certain amount of infrastructure to function, whether or not that infrastructure is located within the nation's boundaries. One of the earliest and most important effects of the post-2001 Afghanistan war was to deprive Al-Qaeda of infrastructure that had served it well as a base of operations and for training and indoctrination camps. The loss did not destroy Al-Qaeda, but it did severely complicate its ability to assemble and train its forces. The attacks on the Al-Qaeda bases in Afghanistan were thus not directly strategic but did lead to more strategic effects as key leaders were killed or forced to flee as communications were disrupted and as training became far more difficult.

Population (ring 4): Nation-states, revolutionists, and organizations such as Al-Qaeda depend on both physical and political support from elements of the population (demographic groups). In the Malayan Emergency—a guerrilla war between the British Commonwealth and the military arm of the Malayan Communist Party—the United Kingdom isolated the ethnic Chinese who drove the insurgency, thus making the emergency manageable.[19] Here a focus on the population center of gravity helped lead to direct strategic results, but not by seeking to kill large numbers of the population indiscriminately, or primarily to win the "hearts and minds" of the relocated people.

Fielded Forces (ring 5): When a nation-state or a group loses some part of its fielded forces, it does one of three things in order of likelihood: organizes and sends more troops, negotiates for time to field new forces or hope for a miracle, or agrees to the proffered peace terms if the terms look cheaper than continuing to fight. Note that the choice rests with the opponent, and that choice is unpredictable. Thus affecting fielded forces is usually a difficult means to a murky and distant end and is rarely directly strategic.

The actual plans associated with each center of gravity are called impact plans, and they have both a strategic and a tactical component.[20] The former describes the required end state for the center of gravity, while the latter lays out the tactics (the actions) needed to reach the end state. Planners at the highest possible level (which ideally means the national command authority or its equivalent in a nonstate situation) should devise the strategic portion of the impact plans, whereas the tactical portion may be developed at lower execution levels. It is essential that the strategic portion of the impact plan be clear and that the organization's leadership understand and approve it. Once the plan is approved, people throughout the organization must translate it into practical steps; no local revisions should be permitted to the strategic portion. Note that strategic impact plans are a senior organization (nation or other) leadership responsibility in which the military, along with other elements of governance, must participate. A strategic impact plan has three components: desired effect (what must happen to a center of gravity, its end state), measures of merit (measures of progress toward the desired effect), and time frame (when the desired effect must be realized).

As in identifying centers of gravity, development of strategic impact plans must carefully avoid getting into tactics. If the planner has identified the center of gravity and can describe a desired effect for it, appropriate tactics almost certainly exist that will realize that effect. An early emphasis on tactics, however, locks the plan into a particular course of action, which is not desirable at a strategic level. Starting with the final step—choosing a tactic, such as a ground attack—subverts the entire strategy process and is highly unlikely to lead to the optimal approach. The strategic approach frees planners to consider and to mix every conceivable way to change a center of gravity—a bribe, an aerial bomb, a hack, a proxy, a conference, an award, assistance funding, or a thousand other possibilities.

Table 3.1 shows some highly abbreviated impact plans for illustration purposes. Note that each center of gravity must have its own impact plan. In the real world, the plans would be more complex, but not excessively so. If planners cannot describe desired effects and measures of merit in three or four bullet points, they have not thought through the issue carefully enough.

Selection of centers of gravity and development of impact plans should start at the highest reasonable level, at least one level above the obvious. Starting at a lower level without consideration for the higher level may often lead to success at the lower

TABLE 3.1 ILLUSTRATIVE IMPACT PLANS FOR CENTERS OF GRAVITY IN AN ENEMY STATE

SAMPLE ENEMY CENTER OF GRAVITY	DESIRED EFFECT	MEASURE OF MERIT	TIMEFRAME
Dictator Abdul (Ring 1)	No longer in power	Deceased/ in exile	One month
Former King (Ring 1)	Accepted as new ruler	Exercising power/ tribal vote majority	Three months
Communications (Ring 2)	Not available to dictator	Orders not received/ activity uncoordinated	Three days
Biological Weapons (Ring 2)	Incapacitated and isolated	Production stops/ no removal of stocks	One day
Military Officers (Ring 4)	Align with former king	Unit defections to king's standard	Two weeks

level while creating potentially fatal results at a higher level. Thus, the top-level strategic plans for an operation in another country should start with global centers of gravity, as doing so will avert major downstream problems and will also highlight opportunities. A good example is the previously cited case of von Schlieffen's excellent plan to defeat the French Army, which failed to address Great Britain as a center of gravity that needed to be kept out of the war.

Plans should also consider the higher level because affecting a parent system will have a significant impact on targets at the lower level. As an example, a plan to neutralize an air base should begin with the enemy system of which the base is a part. Attacks on the parent system might have shut off electricity, logistics computer systems, fuel replenishment, air traffic control, air defense warning and protection, and communication with higher headquarters. The airfield may remain undamaged, but its capability to generate effective sorties will be severely limited. In these circumstances, it may be unnecessary to attack the airfield directly, or the attack could be directed against limited parts of its operation for reasons connected to the future picture for the opponent, perhaps a desire to hold peace talks at this location. In other words, plans should never involve using force against any target before understanding the context of the attack and the system that encompasses the entity being attacked.

A center of gravity at the strategic level is frequently complex and can seem impossible to affect. The solution becomes tractable thanks to the fractal nature of any center of gravity. That is, any center of gravity can be broken down into its own five-ring system, and this process can continue through many nested rings until the plan arrives at an entity that is irreducible for practical purposes. In real-world analysis, planners rarely need more than two iterations to find a component against which to operate to produce the desired strategic effect on the parent center of gravity. This is like examining a malfunctioning joint such as a knee with an x-ray; the idea is to find the specific areas to repair in order to make the knee work again.

The electrical system was a key strategic center of gravity in the first Gulf War, but it was far too complex to understand at the first level of analysis. Breaking it down into its own five-ring system, however, made it easy to find the component to attack to realize the strategic desired effect of denying electricity to the Iraqis during the war, but doing so in such a way as to permit rapid, economical restoration of power after the war, which was part of the postwar future picture for Iraq.

The fractal analysis of the electrical system showed probable leaders such as a government ministers in ring 1, process elements in ring 2, infrastructure such as transmission lines in ring 3, various population groups such as engineers and workers in ring 4, and repair teams and security groups in ring 5. Several features stand out. Planners had at that time lacked detailed information on the electric system leadership group (except for possibly the minister of the interior in ring 1) and expected it would take a long time to find its members and affect them in some way. Although there were many possibilities in rings 3 through 5, none of them looked as though their attack could lead to rapid shutdown of electricity. They realized that trying to affect the electrical system by attacking entities in these rings would entail large costs and yield a low return on investment. As a result, the planners focused their attention on ring 2. It was not necessary to be an expert on electrical systems to realize that it must include three major components: energy conversion processes such as hydroelectric, the generators themselves, and voltage-control processes such as step-up transformers. They realized that damaging the first two elements would almost certainly be very expensive to repair and make it difficult to restore the system, which would violate the future picture for Iraq and the desired system effects. The planners then tentatively chose the voltage-control function (the step-up transformers), which a little research showed would be difficult for Iraq to fix during the war but not when it was over. The planners then asked the intelligence agencies for the detailed information needed to plan attack operations. The fractal approach makes it easy to understand a complex system and also allows planners to direct their intelligence resources against one small point rather than asking for all information about the electrical system in the hope that some useful nuggets will emerge.

During the planning process, fractal examination is usually most appropriate at the campaign level, where campaign teams seek the best way to achieve the desired strategic effect on their assigned centers of gravity. Sometimes it can also prove useful at a strategic level simply to gain a better understanding of the problem. Note also that it is normally necessary to address more than one fractal center of gravity to create the desired effect on the parent center of gravity. That only one was needed in this illustration from the Gulf War was a piece of good fortune that may not repeat itself in other situations.

If we end up choosing to use force as a major or complementary way to achieve strategic objectives, the system methodology just described (or something similar to it) is crucial to the effective exploitation of airpower. This methodology allows us to select the most appropriate centers of gravity and then apply airpower (if appropriate) to produce direct strategic results. The overwhelming advantage of airpower is its ability to operate directly against centers of gravity—targets—that are directly connected with the strategic objectives of the conflict, almost regardless of the nature of the objectives.

The systems methodology helps us avoid the siren lure of "battle," and it keeps us away from starting with the "means" á la Clausewitz while only giving a nod to "other ends" and really having no clear idea exactly where the "means" are going to lead. To the extent that national leaders understand this methodology, they understand the value of airpower; to the extent they do not they will not understand and will be a victim of thousands of years of tactical history that has lost much of its relevance. There is, however, another critical and generally ignored component of strategy that accentuates even more the importance of airpower, and that is time itself, which brings us to the third strategy question: How?

How Much Time Do We Have? (How?)

Leaders of any competitive enterprise, to include the leaders of a nation (or any other group) must understand the importance of time, for it is critical yet frequently mismanaged or even ignored completely. As Sun Tzu said two millennia ago, "Thus, though we have heard of stupid haste in war, cleverness has never been seen associated with long delays. There is no instance of a country having benefited from prolonged warfare."[21] This is as true today as when it was written, except "long" or "prolonged" may have meant many months in Sun Tzu's era whereas today it could mean hours or days. Simply, short is categorically good and long is categorically dangerous and bad. This concept flows from the "time value of action," which in turn derives from the phenomenon of shock effects resulting from compressed, parallel attacks on centers of gravity. The opposite of parallel attack is serial attack, in which attackers attempt to affect one or a small number of strategic centers of gravity sequentially over time.

The way to realize the future picture is via system change, which we do by affecting one or more of the opponent system's centers of gravity. The impact on the system as a result of affecting the centers of gravity will be a function of how quickly they are affected. Affect them too slowly (serially) and the system will probably find ways to repair itself, protect itself against further attacks, and begin its own operations against its opponent's systems. Conversely, if enough centers of gravity are affected quickly enough (in parallel), the system will go into a state of paralysis in which it cannot repair itself, protect itself against future attacks, or make competent attacks on its opponent's systems.

When we go to war, we want to have as high a probability of success as possible (and at the lowest possible cost). Our probability of success is a function of what we do to the enemy and the time period in which it is done. The following equation indicates that the probability of success in changing an enemy goes up as we lower the opponent's energy (by affecting its system through attacks on centers of gravity) and decrease the time needed to do it:

$$P_S = \Delta(Energy)/Time$$

Whether planners pursue the centers of gravity serially or in parallel is critical and has a huge impact on the outcome of the conflict. Serial operations attempt to affect one or a small number of strategic centers of gravity sequentially, one at a time. In the past this generally meant focusing on only enemy fielded forces. If they succeeded, operations then moved on to the next problem, and so forth. Serial operations give an opponent ample opportunity to react, which in turn confronts the attacker with an entirely new set of problems. By contrast, parallel operations focus resources on changing an entire system at once, whether that system is a market, an organization, a military unit, or a nation-state opponent. This concept of parallel operations was designed to produce significant effects very quickly with minimal cost and risk and to create change that would last, yet the concept is not widely understood or used.

Great military commanders have always understood the importance of time. Indeed, moving quickly to the right place distinguished the operations of Alexander the Great, Napoleon, and George Kenney. In the American Civil War, one of the Southern commanders, Nathan Bedford Forrest, is still remembered for his succinct overview of his own success: "Get there first with the most." Despite this general understanding of the time element, until recently combatants have found it difficult to do much other than seek to move their military units (of any size) a little faster than their opponents could move their own forces. In general, the speed differential was small, because both sides usually had similar technology for mobility.

Parallel war against an enemy system is completely different from the old serial concepts of war mostly conducted as force-versus-force operations. Indeed, the ability to conduct parallel war is one of the major reasons why the very nature of war has changed in a revolutionary way. A parallel approach to warfare enables planners to capitalize on the value of time and, most important, understand how to make it their servant.

The last half-century or so offers several examples of both the serial and parallel approaches. In World War II, American aerial attacks on German targets in 1943 were serial; the U.S. Eighth Air Force, for example, only hit about eleven target categories that could be considered "centers of gravity," and six of these were directly or indirectly against fielded forces (aircraft and ships).[22] Of the remaining five, only the attacks on the marshaling yards, the synthetic oil installations (three attacks total against two locations), and to some extent the ball bearing production approached being a second ring (processes) center of gravity that could have had a general impact on Germany as a whole. Note that there were no attacks on ring 1, leadership, and no attacks on such key ring 2 (processes) targets as electricity, command and control communications, energy other than oil, transportation other than rail marshaling yards, food, finance, or radio broadcast, to name just a few. At the time, of course, some of these centers of gravity were beyond the available technology. In addition, the rate of attack was measured: there were no attacks (defined as more than ten aircraft participating) in twenty-one weeks of the year and the median number of attacks per week for the entire year was just *one.*[23] Although these attacks did substantial damage and forced the Germans to reallocate resources for defense and repair, Germany as a system was functioning well at the end of the year. For reasons of weather and bomber diversion to support the planned cross-Channel invasion, it was not until the end of 1944 that attack intensity effectively moved operations from serial to parallel. By the end of the war in May 1945, Germany was in a state of paralysis because too many things were broken to allow effective repair, defense improvement, or competent counterattack.

A similar pattern occurred in Operation Allied Force against Yugoslavia (Serbia) in 1999. Attacks in the first month were serial and directed largely against fifth-ring fielded forces. Serbian leader Slobodan Milosevic's forces could still operate effectively under this attack methodology and even stepped up their operations in Kosovo. Once NATO shifted to parallel attacks in the second month and included direct leadership and process centers of gravity, internal dissension at the highest levels of government appeared within a week. Two weeks later, Yugoslavia claimed it was withdrawing forces from Kosovo, and in the eighth week following the change in attack methodology Yugoslavia essentially offered to capitulate by saying it would accept the European Group of Eight "principles for peace."[24]

The "time value of action" is of tremendous importance. As attacks shift from the parallel domain to the serial domain, their probability of success begins to fall dramatically. If an operation takes a long time, the chances become very low indeed.[25] This applies to both sides, although their respective centers of gravity may be much different. Since good strategy depends on understanding probabilities, deliberately embarking on a long, serial war is counterproductive.

To the extent it is possible to bring the majority of relevant centers of gravity under attack in a compressed time frame, the attacked system will find it almost impossible to react in any meaningful way. For example, an individual combatant can continue to fight despite a serious wound to a hand or a foot, but someone who simultaneously loses both feet, both hands, both eyes, both ears, and his voice would be in a state of paralysis and unable to fight. In war, planners want to make the equivalent happen to the system they wish to change. The more compressed the operations, the higher the probability of success.

No plan ever has P_S of 1 (100 percent probability of success), but a well-conceived and well-executed plan may come very close. As operations move from the parallel domain into the serial domain, P_S begins to fall quickly toward but never reaching a P_S of zero. Nations and groups have won long, serial wars, but it is important to emphasize that the probability of doing so declines dramatically with the length of the conflict.

Planners might then think that if their own probability of success in long war is low, the opponent's chances are high, but the time value of action curve applies to both sides in a conflict. Although each side has different centers of gravity, the level of time compression in the attack determines the probability of success.

As an illustration, think about the Vietnam War. From the American standpoint, there were a number of centers of gravity that had to be successfully attacked in order to win the war. In North Vietnam, they included Ho Chi Minh and his lieutenants; the Communist Party leadership; internal and external communications; military and civilian provisioning (internally or externally provided); energy systems including electricity and petroleum; mobility infrastructure such as roads, railroads, and bridges; population groups; and fielded forces (offensive and defensive). Had the United States been able to bring all of these under parallel attack in 1965, North Vietnam would have been unable to prosecute the war. In South Vietnam, centers of gravity relevant to American success included South Vietnamese leadership; economic and financial systems; provisioning; mobility infrastructure; population groups; South Vietnamese fielded forces; and the Viet Cong (which had its own centers of gravity). If North Vietnam had been out of the picture and if the centers of gravity in South Vietnam had been brought under compressed attack, the United States would probably have achieved its objectives in that area. In actuality, however, the United States did not attack most of the centers of gravity in either North or South Vietnam and the ones it did attack were struck serially. If an observer had known in advance the serial nature of the attack, he would have noted that American operations were on the right side of the graph with correspondingly low probability of success (but not zero).

Now consider the situation from the standpoint of the North Vietnamese and Viet Cong. From their perspective, the American centers of gravity that they had any hope of affecting included the president of the United States; the U.S. Congress; key American media (print and television); economic system; population groups; and American fielded forces operating in or over South and North Vietnam. Their centers of gravity in South Vietnam included South Vietnam leaders; economic system; provisioning; mobility infrastructure; population groups; and fielded South Vietnamese military forces. If the North Vietnamese and the Viet Cong had been able to have an appropriate effect on most of the American centers of gravity, their probability of success would have been high. To the extent that their operations were serial an observer would have predicted a low probability of success (but not zero).

In the Vietnam War case, the United States had the ability to operate in parallel but chose not to do so for a variety of reasons including psychologically based ideas such as escalation and coercion theory. On the other hand, the North Vietnamese and the Viet Cong had no way to conduct a similar parallel attack on American centers of gravity. To be successful, they had to hope that their serial efforts would produce a favorable conclusion before the United States decided to make the war parallel. From the American standpoint, there was some possibility that its serial operations could also lead to success. In both cases, however, the probabilities were low. The difference was that for the United States, serial operations provided a high-risk (low P_S), low-reward situation whereas for the North Vietnamese and the Viet Cong, their serial operations presented a high-risk (low P_S), high-reward situation. Thus, a long war was a high-risk, costly proposition for both sides

Another phenomenon occurs as planners move into the serial domain in war or business. In business, costs include time to market, inefficient use of people and facilities, and lack of strategic information. In war, the cost of serial operations—in lives, money, and equipment—rises dramatically for both sides. Paradoxically, parallel attacks are actually less costly for both sides, although such attacks may call for higher initial commitment and expenditures. The enormous difference manifests itself when planners examine the cost from inception to conclusion. That is, a parallel effort will require more resources initially, but because parallel attack leads to a much shorter conflict, the overall cost of the war is much lower than in a long serial war. Thus the long, serial Vietnam War cost about eight times as much in dollars as the short parallel first Gulf War, and about two hundred times as many Americans killed in action. In addition, the cost associated with operating in the parallel domain is relatively predictable, in part because it is far easier to make predictions for the short term than for the long term. Because of this, forecasting the cost of serial operations is extraordinarily difficult and the actual costs almost always far exceed the estimates; abundant examples include government acquisition projects and the expected cost of wars.

Simply stated, whether in war or business, the traditional approach to the time element is exactly the opposite of the desirable one: managers and planners typically ask themselves how long something *will* take rather than deciding how long it *should* take in order to create parallel effects and succeed at an acceptable cost. The right strategic question is always: "How much time do we have?" So important is this concept that national leaders can use it to help decide whether or not they want to go to war at all. If they are unable or unwilling to operate in the parallel domain, their first choice should be to look for ways to avoid war, probably a reasonable course in most instances. The United States would have saved many lives and many dollars had it applied this concept in 1963 before committing itself to such an expensive undertaking.

This discussion shows why planners should avoid serial operations whenever possible, and it is almost always possible to avoid them. The plan should not give the enemy system the opportunity to pursue its own goals, to repair itself, to figure out how to thwart the attacker's next move, or to counterattack.

Parallel operations, whether in business or elsewhere, are faster, cheaper, and more likely to achieve success. They do, however, require a different mindset and an organizational structure that may be much more dynamic than that of most organizations.

Many planners believe that parallel operations will demand more information and cost more than serial ones. Paradoxically, exactly the opposite is the case. For example, serial operations demand that the attacker have accurate information about each target. If forces attack the wrong place at the wrong time, and that operation is the only one occurring at that time, the entire campaign comes to a halt. Serial operations depend on achieving their goals before forces can proceed to the next objective. In contrast, parallel operations do not demand perfect knowledge or perfect execution against any given target, because they seek to achieve a systemic effect rather than the single-point effect that characterizes the serial world. From a cost standpoint, planners should estimate the likely cost of a long serial operation. On seeing this sum, they should realize that a small fraction committed up front could move operations into the parallel domain where costs will almost certainly be dramatically lower.

The other big advantage of parallel operations is the ability to decide quickly that there has been a strategic error in either planning or execution. Movement into the serial domain is one of the major signals that indicates serious problems and the need to do something different and do it fast, whether that is a major change in plans or actual abandonment of the war or project.

A good commander will have a smart future picture and a plan to attack the opponent's system and centers of gravity, and commanders will be confident in their ability to operate within the available time via parallel operations. Nevertheless, if commanders do not also know how to deal with both successes and failures in their plans, they are not strategists, which leads to the last of the key strategy questions: How to exit?

Transitioning from Success or Failure (Exit?)

So far, this chapter has considered the Where, What, and How of strategy. The fourth central strategy question concerns "Exit" and is particularly difficult to address. Nations, wars, weapons, technologies, units, and all other entities have come to an end since the dawn of human history, but planners rarely take the inevitable end into account. The failure to plan well (or at all) for inevitable end points at best leads to wasted resources and opportunities, and at worst to disaster. As counterintuitive as it might seem, end points tend to be the most dangerous and expensive aspects of strategy. Initiating an action is rarely fatal, but the failure to plan for disengagement or escape almost always leads to catastrophe.

The term "Exit" nicely captures the concept but requires a little thought to understand why. If someone makes a successful purchase in a store, the next step is to "exit" and move on to something new. Likewise, if the "someone" fails to find the sought-after item, the next step is also to "exit," perhaps to go to another store or perhaps to give up the quest entirely. In both cases, "exit" connotes making a strategic move in response to success or to failure. What would make no sense in the first case would be to stay in the store and continue to buy things merely because of a good purchase, regardless of what had been sought. Likewise, it would be less than bright to camp out in the store in the hope that the desired item might show up at some point in the future.

As human beings, we do a poor job of "exiting." After success, the tendency is to keep doing the same thing because it worked so well; and the tendency when we fail is to deny error and continue doing the same thing because abandoning it is too painful. Of all the things that good commanders and strategists must do, "exiting" is the most difficult, and most often botched.

A few examples from history will illustrate the need for exit scenarios from both success and failure. Hannibal and Carthage had but one objective in the Second Punic War: the complete overthrow of Rome. Their efforts progressed well through Hannibal's spectacular success at Cannae, but shortly thereafter it should have become apparent that the chances of defeating Rome completely were small at best. If Carthage's leaders had established end-game plans while Hannibal's army was still superior in numbers to the Roman armies on the Italian peninsula, they almost certainly could have arranged a negotiated settlement or organized an orderly withdrawal. Instead, the situation eventually reversed itself to the point where a Roman army ventured to Africa, defeated the Carthaginian forces, and put Carthage in a perilous position from which it was never able to recover.

In 1905, General Alfred von Schlieffen designed a plan for a war with France that depended on a quick victory. When the German attack in August 1914 stalled, rather than admit that the premises on which the campaign had been based no longer existed, the Germans decided to persist with an operation that would have been soundly rejected had it been suggested before the war. The results were four years of static attrition warfare, the eventual destruction of the German monarchy, and the stage set for still another disastrous war.

In 1944, the Germans made a desperate attack on the American lines in the Ardennes. One of the key objectives was to capture St. Vith within twenty-four hours, since the rail center there was essential to supplying two German armies. But a valiant Allied defense held out until the fourth day, and Germany had no plans to deal with a failure other than pressing on with the original plan. With no rail center, the Germans soon experienced a shortage of supplies, especially fuel and ammunition, which made it impossible to attain their other objectives. Had the Germans devised an agreed exit plan, the armies the Germans lost would have been available to slow the Soviet advance to the west and the postwar world might have been far different.

Some twenty years after its agonizing experience in Indochina, the United States sent forces to Somalia to break the internal blockades set up by various warlords, which were causing widespread hunger and starvation. The mission succeeded at low cost to both sides. Again, however, the United States had failed to formulate an agreed exit plan and drifted into expanding the mission to one of bringing democracy to Somalia. The aim may have been noble, but U.S. decision makers had not thought through the methods for achieving it and did not assign the forces appropriate to the new operation. The result was the predictable Battle of Mogadishu, made infamous in the book and movie *Black Hawk Down*, and the resultant public pressure for a precipitous withdrawal.

Exiting not only constitutes the most dangerous phase of strategic operations, but it is also the most difficult to plan and execute. Part of the problem lies in the human tendency to assume (despite all the evidence) that the next operation will have no flaws or that the commanders will be so capable that they can adjust to changing circumstances on the fly. Unfortunately, carrying out a good exit becomes nearly impossible unless the exit conditions are planned and agreed before operations begin. Defining exit points for both success and failure requires careful analysis, but if planners do not determine them in advance (and take these plans as seriously as all others in the campaign), they will find many reasons why the force should "give it just one more chance."

System Warfare and Airpower: Exploiting the Revolution

This discussion of airpower and strategy began by suggesting that our concept of war and accompanying vocabulary were outmoded and dysfunctional and were still based on an ancient idea of war highlighted by Clausewitz's focus on battle. Old ideas are not necessarily bad ideas; Aristotle's insights into philosophy are as valid today as they were two thousand years ago. On the other hand, Aristotle's concept of a geocentric cosmos gave way to Copernicus' heliocentric cosmos, which in turn has given way to a variety of post-Einsteinian views. Thus we should be quite comfortable moving beyond the old concepts of war.

The old ideas had some practical value in the past, when the military forces available to any state or organization were small and their speed and range were limited. If an organization defeated the military of another organization in battle, chances were good that nothing stood between the victor and the real reason for war: seizing wealth in the form of crops, land, gold, or slaves. On the other hand, failure to overcome the opponent's military meant that your own wealth lay open to seizure and destruction. Most of our thinking and operations, then, really flowed from the extraordinarily limited capability of the available forces, so there was no compelling reason to think beyond the battle.

Imagine, however, that armies of old could have instantly transported themselves into the rich heartlands of their opponents, where the plunder would have been theirs for the taking. Would our whole concept of war not have been much different? In addition, the military forces themselves could rarely attack more than one thing at a time, so they had to proceed serially. It is only within the past seventy-five years that technology has made it possible to attack multiple centers of gravity in parallel, to strike these multiple centers of gravity in compressed time frames, and to control the damage done to them. The nature of war itself has changed. Can there be any question that we desperately need to rethink the concepts and vocabulary associated with it?

Airpower enables us to think about conflict from a future-back, end-game-first perspective as opposed to a perspective based on the serial battle obsession of Clausewitz and his followers. It also allows us to get closer to our goal of making every act of war as strategic as possible by carefully matching what we hit, the damage we do, and the casualties we create to the peace we want to create: our future picture.

When we engage in conflict, our strategic objectives should always be to create a better peace. A normal part of a better peace is one in which the vanquished do not bear such hatred for the victors that another trial is inevitable. As Montesquieu suggested, . . . nations should do each other the most good during peacetime and the least harm during wartime without harming their true interests, if peace is to be anything more than a temporary suspension of arms."

One of the ways to reduce postconflict enmity is to reduce the suffering and recovery time of the defeated party, which means in part avoiding traditional wars with their perverse and long-lasting impacts. To do so, it is imperative to ensure that all damage and destruction are inflicted in connection with a specific future-picture, postwar, win-the-peace objective, and that things are damaged or destroyed only to the extent absolutely required.

Airpower already has the ability to deliver energy with great accuracy (precision of impact). In the world of the early twenty-first century, the majority of weapons employed by high technology air forces hit their intended targets. There is, however, another "precision" and that is precision of effect. Precision of effect means that a weapon does only the damage to a target that is necessary and intended. Thus, if the intent is to affect a communications network by destroying a switch in a basement, a weapon with precision of effect does that and only that; the building and the people in it are unharmed while the switch disappears. The new small diameter bombs are an interesting initial step in this direction.

With precision of effect combined with precision of impact, war matched to the desired peace comes closer to reality as it becomes possible to affect those things that are necessary to destroy (or paralyze for a period of time) without doing unneeded damage or causing unplanned casualties that would militate against a successful peace and that waste resources in the process.

Time in war is critical. Short is good, long is bad, and short time frames are essential to create system paralysis through parallel war operations. Not only, however, does striking relevant centers of gravity in a compressed time frame significantly increase the probability of success, but also it also dramatically reduces the cost of the war, to both sides. Long wars are incredibly expensive, and long, traditional wars based on battles cut wide swathes of death and destruction to soldier and civilian alike in the areas in which they are fought. Airpower provides the technical means to make wars short and thereby significantly reduce the death and destruction visited on the attacked side. The first Gulf War provides a good example where not only were losses of the victor historically low, but so also were the losses suffered by the vanquished.

Simply put, airpower allows us to wage war in ways that were unthinkable even a few years ago. It is following the classic trajectory of a disruptive innovation, which is from a substitute for the old to completely new concepts of operation. It works, as is clear from the wars since 1991 where it was the primary force.

Before beginning the discussion of how to increase understanding and employment of airpower, let us summarize the key points so far in this chapter. Most important, airpower can only realize its potential of moving us into a new age of warfare if it is tightly linked with a future-back, end-game strategy that rejects anachronistic ideas about battle-centric war. This means developing a strategy for a conflict that starts with a future picture, determines the systems and centers of gravity that must change to realize the future picture, takes into account the impact of time, and preplans for exits. In selecting the right centers of gravity, the focus should be on direct strategic centers of gravity to the maximum extent possible. Time matters; fast action and short conflicts are imperative and far less expensive than slow and long ones. Exit plans and operations are essential and must be integral parts of airpower thinking. Finally, it is essential to remember that the objective of a conflict is to achieve a future picture (win the peace), not to fight battles, and not to kill and destroy. As we consider conflict, we should do everything possible to develop a strategic war plan based on systems and centers of gravity. And we must be unrelenting in opposing the inevitable hysteresis effect that tries to make us revert to old-style war and battle, for "battle" is at best an expensive and risky means to a distant end and should be avoided.

Choosing the Right Tool

A military leader's options for applying force, in the broadest sense, consist of ground power, sea power, and airpower (any of which could be involved with cyber operations). Admittedly, in the world of real organizations, armies and navies have airpower, while air forces normally have little ground power beyond that needed for light security. For the sake of simplicity, this discussion does not reflect the structure of current service organizations, but instead defines "airpower" to mean the use of any guided object that flies through the air and space regardless of who owns it or its launch platform, sea power to mean the use of resources that operate on or under water (but does not include aircraft or missiles launched from ships), and ground power to mean the use of resources essentially tethered directly to the earth, including people, tanks, and artillery. Only after reaching conclusions as to the type of power to employ should military leaders designate the organizations that should own and operate the three types of power.

Ground power, the oldest and historically most prevalent tool of conflict, moves slowly and can normally affect only enemy fielded forces—the outermost ring of an opponent—which are rarely connected directly to a strategic objective. Ground forces have only minimal ability to conduct parallel operations on their own and to operate without significant destruction and bloodshed.

Sea power can operate against centers of gravity directly or closely related to strategic objectives, but only if those centers of gravity are accessible by water, a category that includes many but not all such centers. Sea power can move faster than ground power and can bring more centers of gravity under attack, but in most circumstances cannot execute parallel operations. It has the ability to conduct operations with far less damage to people and infrastructure than does ground power.

Airpower can operate against virtually all the centers of gravity directly related to strategic objectives regardless of their location. It can bring many under attack in compressed periods of time and therefore is well-suited for parallel operations. Finally, it can achieve appropriate effects with relatively little destruction and carnage.

Furthermore, planning for and executing the "Exit" step becomes much easier when a campaign involves air-based (and even sea-based) forces than land forces, since to some degree exit from an airpower war means simply canceling the next day's planned sorties. In reality, many other considerations arise, but the mechanical aspects are quite easy compared to withdrawing large numbers of troops, local collaborators, and equipment from deep within foreign territory.

Airpower has clear advantages and should be the go-to choice for national leaders, but it will not be until those leaders understand it and advocate it, which brings us to the last section of this chapter.

The Strategy of Airpower Advocacy

To this point, we have tried to make the case that airpower can only realize its potential for achieving system-level goals if it is tightly linked with future-back, end-game strategy that rejects anachronistic ideas about war. Airpower enables planners to approach the goal of making every act of war as strategically significant as possible by carefully relating the targets hit, the damage created, and the casualties caused to the peace they want to create, their future picture.

The overwhelming, decisive value of airpower should be clear, but it generally remains unrecognized by the majority of government officials and military officers, including many who operate some facet of airpower. Convincing decision makers of the value of airpower must start by connecting it uniquely to the new approach to success in conflict discussed in this chapter. Air strategists must stop believing that they can ensure proper use of this valuable resource by relying on the two methodologies most prominent in the past few years: trumpeting spectacular technology and asking merely for recognition as equal partners with ground and sea power in the "joint arena." Businesses long ago learned that selling a product involves far more than touting its technical qualities; products sell when customers see them as filling a real need in their lives. To achieve such recognition for airpower, its advocates must highlight its advantages and show convincingly that it fills a vital need better than its alternatives. If the "customers" accept the approach to strategy described in this chapter, airpower becomes the obvious solution; otherwise, decision makers will continue to view airpower advocates as simply hawkers of new and highly expensive technology.

Marketing, then, becomes the top priority for airpower even though few airpower advocates practice it effectively. Those who understand and believe in airpower must direct their efforts toward taxpayers and decision makers at large and indeed must analyze the problem in the same way planners analyzed problems such as the 1990 Iraqi invasion of Kuwait. That is, they must have a future picture about airpower, understand the need to change the existing concepts of war within their defense communities, apply their efforts against centers of gravity within that system, and strive to operate in parallel so as to gain maximum probabilities of success at the lowest possible cost. Otherwise, airpower advocates will limit themselves to trying to convince ground power and sea power advocates of something they view as against their best interests.

Success with airpower over the past century has flowed primarily from the former approach, not from the latter. Historically, when the public and senior civilians in government understood the value of airpower—to include the cost of depending on other means—plans for novel application won acceptance. The British use of airpower in 1920s' Mesopotamia, emphasis on airpower in the 1930s as another European war loomed, long-range aerial attacks on Germany and Japan as a major part of the World War II effort, the huge investment in airpower as a weapon and deterrent in the first half of the Cold War, and the use of airpower in the 1990s were possible because of support from the public and senior civilian leaders. None of these would have happened if they had depended on a vote by the "joint team." In other words, airpower advocates have succeeded when they played what we might call the outside game and far less so when they tried to play the inside game. It is simply imperative that airpower advocates make a compelling case to senior civilian leaders and to influential members of the public.

Airpower advocates not only need to connect airpower directly to strategy and market their product well, but they also need to start believing in it. Those who begin a discussion by noting that airpower alone "cannot do everything" do themselves and their listeners a disservice. Undoubtedly, military power alone cannot do everything or even contribute significantly to many national objectives, but if a problem is amenable to military solution, airpower should not be disqualified from any aspect of it. Advocates should begin with "the nonlimits of airpower" in mind: the presumption that airpower can accomplish any military task. Otherwise, advocates create a self-fulfilling prophecy by failing to examine the possibilities for using airpower to achieve a given purpose. Careful consideration of a problem may reveal that airpower is the wrong tool at the moment, but airpower theorists and planners should then consider how to overcome current limitations and hone airpower for future purposes.

Espousing the nonlimits concept of airpower exposes advocates to charges of being airpower zealots, of failing to consider the "joint" dimension, or parochialism. But airpower advocates need to become confident enough to shrug off these labels and recognize them as a particularly low form of ad hominem argument that indicates a lack of substance or logic with which to counter the airpower proposition. Decades ago, airpower advocates refused to be marginalized by such attacks and pressed on to achieve the impossible time after time. To attain better airpower tomorrow and a brighter, more affordable, more effective, and lower-risk future for our nation, they must reclaim the courage and confidence of their forebears. If they can do so, they can reforge airpower into a concept invaluable to their nation and civilization, and one that will return huge dividends on the human and monetary investments needed to realize its extraordinary promise.

If airpower is to achieve this potential, the operators of airpower must understand, believe, and teach end-game strategy as the foundation of airpower. This is the very essence of airpower and where the discussion must start. The technology is extraordinary and beguiling, but is only one of many possible ways to fight wars. When tied to strategy, however, the airpower case becomes compelling because it offers a high probability of success with risks dramatically lower than those associated with Clausewitzian battle dogma. As stated at the beginning of this chapter, airpower advocates must stop trying to use airpower as a substitute for its military predecessors, must connect airpower directly to strategic end games, must adopt a new vocabulary to match its promise, and must become serious promoters not of machines, but of ideas.

CHAPTER FOUR

Fifth-Generation Strategy

Alan Stephens

The truism that insanity is doing the same thing repeatedly and expecting a different result could be applied to the United States' preferred military strategy for the past half-century. Four times American-led armies have invaded and occupied foreign countries, and four times that land-centric model has failed, in Vietnam, Iraq (twice), and Afghanistan.

Remarkably, at the same time, in campaigns that have received insufficient thought from the media and most academies, Western and allied air forces have utterly dominated six major air actions—that is, actions that did not require invasion and occupation—in Iraq (twice), Afghanistan, the former Republic of Yugoslavia (twice), and Libya. It is notable that almost no Western combat land forces were committed to the latter three conflicts.

Western defense forces have been exceptionally good at modernizing their equipment and training, achievements that have produced formidable traditional warfighting organizations. It is debatable, however, whether that modernization has been extended to strategic thinking. Indeed, given the evidence of fifty years of defeat, it is fair to suggest that Western military staffs and academies have been fortresses of obsolete ideas rather than agents of strategic innovation.

The central question is this: Why has the West, led by the United States, persisted in invading and occupying foreign territories—in attempting to "seize and hold ground"—as its preferred military strategy for fifty years when, as the evidence clearly shows, the model is broken? This chapter examines the background to that failure of strategic thought and presents an alternative based on the West's demonstrated single greatest military comparative advantage, namely, airpower.

A Broken Model

The failure of contemporary Western strategic thinking has two main causes. First, simply put, the era has gone in which predominantly white, predominantly European, predominantly Christian armies could stampede around the world invading countries their governments either do not like or want to change. In the global village of the twenty-first century that kind of mentality is obsolete. Countries and interest groups connected by instantaneous communications, travel, trade, finance, and shared individual (as opposed to national) interests no longer accept the assumption of Western superiority that shaped the preceding five hundred years of the international order.[1] Today, for example, the West's "expeditionary war" is someone else's "invasion." The shift in terminology is both instructive and profound.

Second, the subjects of occupation have learned how to exact costs that are too high for liberal-democratic societies to bear in situations that do not represent a threat to their national survival. The kinds of casualty rates accepted so carelessly by British, French, and German generals in World War I would be intolerable today. U.S.-led armies might not have been defeated on the battlefield in the traditional sense in Vietnam, Iraq, and Afghanistan, but, as body bags continued to be sent home over an extended period, they were conclusively beaten on the field of public opinion. Simultaneously, the economics of occupation have became an own center of gravity for invaders. For instance, no Western general can plausibly explain why the most technologically advanced armies the world has ever known should have to spend hundreds of millions of dollars trying (unsuccessfully) to counter improvised explosive devices—cheap, homemade bombs used by socially primitive opponents.

To put it another way, in each of those "expeditionary" wars, the West's apparently technologically and sociologically inferior opponents have been able to define the nature of the fighting, a situation that seems to defy logic—well, habitual military logic anyway.

Strategy must be the starting point for any analysis of those failed campaigns. Three conspicuous common features emerge. All were land-centric, all involved large armies, and all began with the invasion and occupation of a foreign country. Those features imply a Clausewitzian foundation to the campaigns' conceptual origins.[2]

First-Generation Strategy

Almost two hundred years after Carl von Clausewitz's death, the celebrated Prussian's work remains the subject of regular disagreement, a phenomenon that might in itself suggest fractures in its contemporary relevance.[3]

Let there be no doubt: Clausewitz is one of the most important philosophers of warfare. His insights into the psychology and political dimension of war are both peerless and timeless. Nowhere is this better revealed than in his notion of the "trinity," a set of interactive forces that impel the events of war. Consisting of primordial violence, the play of chance and probability, and the function of war as an instrument of policy (that is, as an action that is not an end in itself but that must contribute to some superseding political purpose), the trinity's implications for a particular geostrategic setting will be present in any decent campaign plan.[4]

At the same time, even Clausewitz's staunchest supporters acknowledge the prominence throughout his opus of such dubious warfighting concepts for the twenty-first century as mass (as in very large numbers), taking and holding ground as an end in itself, close engagement, and attrition.[5] Furthermore, although his analysis contains commentary on the spectrum of conflict from absolute war to limited war, and also on counterinsurgency warfare and the possibility of fighting to achieve limited political objectives, in the broader sweep of his work those are second-order considerations. There is a consensus that the primary influence on Clausewitz's thinking was the large-scale interstate conflict typical of the Napoleonic Wars, which had "an indelible impact . . . leading him to emphasize the centrality of the bloody and decisive clash of arms throughout much of his career."[6]

Clausewitz's opinion was necessarily influenced by the technology of his era, which among other things imposed strict limits on information flows, rates of maneuver, and precision firepower. In that context it was understandable for him to emphasize the massed, bloody clash of arms characteristic of a Napoleonic battlefield. But today, when technology has fundamentally transformed those capabilities (and much more besides), it is irrational to cling to forms of warfare shaped by early-nineteenth-century circumstances, to models we might describe as "first-generation" strategy.

The Clausewitzian model was exposed in its crudest form one hundred years ago on the Western Front in World War I, when military strategy, such as it was, consisted essentially of massive artillery bombardments, vast numbers of infantry facing off across static trench systems, frontal assaults into withering machine-gun fire, merciless attrition, and immense casualties.

Twenty years later in World War II, grinding massed clashes again characterized much of the fighting, especially between the Soviets and the Germans on the Eastern Front (where arguably the war was won and lost). Still, some intellectual progress was evident, especially in the deserts of North Africa and the maritime reaches of the Pacific, where a more fluid approach to maneuver and the application of firepower was introduced via mechanized transport and air and naval forces.[7] Similar progress was evident early in the Korean War, when in September 1950 the UN commander Gen. Douglas MacArthur mounted an audacious amphibious landing behind the enemy lines at Inchon. But from mid-1951 onwards the fighting degenerated into a First World War–style standoff across the 38th parallel, the generals seemingly having run out of ideas.

Even when the attention of Western armies began to turn toward the apparently more agile theory of counterinsurgency warfare in the second half of the twentieth century, their thinking was unable to break free from the first-generation dogma of seizing and holding ground, mass, and attrition.

During the occupation of Vietnam from 1962 to 1973, for example, American-led forces under the pedestrian leadership of Gen. William C. Westmoreland eventually adopted a strategy of "search and destroy" against the nationalist/communist army (consisting of Viet Cong and North Vietnamese soldiers) in South Vietnam.[8] Search and destroy amounted to a series of limited duration, large-scale operations into enemy-dominated territory, mounted from vast base camps. The intention was to demolish enemy strongholds, capture supplies, cut lines of communication, encourage surrender, and kill troops and their supporters.

Despite search and destroy's ostensibly mobile nature, one of its defining characteristics was Westmoreland's fixation on "mass," reflected in his regular requests for more soldiers, then more again. Between 1962 and 1968 American troop numbers in Vietnam increased from 700 to 540,000.[9]

Against that background, it should not be surprising that statistics became both a measure of achievement and a de facto strategy. Numbers became ends in themselves as daily lists of "achievements" were reported back to Washington from Saigon: so many bridges destroyed, so many food and weapons caches captured, so many villages pacified, and, most perniciously, so many enemy troops killed in action. It was almost inevitable that ambitious field commanders would inflate the figures in their reports. Worse still, those figures, especially the daily reporting of communists killed, came to be seen as an indicator of progress. Counting body bags became the strategy.

The Clausewitzian obsession with mass as in "very large numbers" was again evident during the invasion of Iraq in 2003, a campaign that eventually destabilized every country in the Middle East and which left Shia-dominated Iran and Hezbollah with little option other than to fight against the West and its interests in Syria, Lebanon, Egypt, Israel, Jordan, and the Gaza.[10] The case in point this time was the so-called "surge" of U.S. land forces in mid-2007, a strategy devised in response to the deteriorating situation on the ground.[11]

Orchestrated by the then-commander, Gen. David Petraeus, the surge has been widely accepted as some kind of strategic triumph. This perception, which for obvious reasons suited the administration of U.S. president George W. Bush and his senior officials, has become conventional wisdom. However, the perception was never adequately tested in the popular press, and it seems probable that any improvement on the ground in Iraq at the time of the surge was a consequence more of fortunate timing than of masterful strategy.

Two events were critical. First, the initial reduction in the level of violence that accompanied the surge seems to have owed more to a concurrent policy of paying bribes to Sunni insurgents than to the arrival of more American troops.[12] And second, the surge coincided with a dramatic about-face by an important group of Iraqi guerrillas.[13]

Known as the Sunni Awakening Movement or the Sons of Iraq, this group from Iraq's largest province of al-Anbar had loathed the dictator Saddam Hussein, but they loathed the Western invaders even more and had been fighting against the Americans for years. By 2007, however, the people of al-Anbar had become alienated by the vile behavior of their Al-Qaeda allies. Led by a prominent Sunni insurgent, Sheikh Abdel Jabbar, the awakening brokered a cease-fire with the regional U.S. commander and then mobilized thousands of their fighters to crush Al-Qaeda. According to Jabbar, "It was the Iraqi insurgents who switched sides to fight al-Qaeda who primarily brought about a more peaceful Iraq. . . . The tribes in Anbar are the ones who defeated terrorism. . . . As for the Americans, they didn't fight al-Qaeda, and so the increase in their numbers didn't have an effect."[14] It says something about the defense policy debate in the West that Sheikh Jabbar's claim, which effectively refutes the received wisdom of the surge as strategy, has gone largely unremarked in the media and defense/academic circles.

Regardless of any military outcomes that might or might not have been associated with the surge, the fact is, seven years later Iraq remains a society defined by political instability, extreme levels of violence, an economy in tatters, and a culture of endemic corruption. Worse still, it is increasingly anti-Western. In short, none of the fundamental problems that plagued Iraq before the invasion and the surge have been answered.

The Emperor's New Clothes: Counterfeit COIN

Even though first-generation strategy had failed in Vietnam and Iraq, its defining features were once again evident during the U.S.-led campaign in Afghanistan that started in 2001 and continues at the time of writing (early 2014). That campaign, incidentally, began with an exceptionally successful operation involving advanced airpower, substantial numbers of indigenous ground forces, and small numbers of Western special forces who advised the local militias and provided targeting information for air strikes. Characterized by knowledge dominance, speed, fluid movement, precision, and a fleeting footprint, the operation was the antithesis of first-generation strategy. Within ten weeks it had routed the Taliban and Al-Qaeda. It was only when the United States invaded and occupied Afghanistan that the campaign unraveled.

This time the invading army's declaratory strategy was the fashionable notion of counterinsurgency warfare, COIN in the military vernacular. COIN was by no means a new concept, but its proponents asserted that a "modern application" would be ideally suited to Afghanistan, a society that in many aspects could be regarded as medieval. In fact, this latest iteration of COIN turned out to be nothing more than just another case of the emperor's new clothes, a theory lacking intellectual credibility and, worse still, common sense.

General Petraeus was again the U.S. commander during a critical stage of the occupation in 2010–11. Described as an expert in COIN, Petraeus at the time enjoyed enormous popular and political support, even though he was representative of a generation of Western army officers whose experience in the field was one of repeated failure. The irony continued. Petraeus was also a principal author of the American army's latest manual of counterinsurgency warfare, FM 3-24, a booklet that had acquired near-mythic status among Washington insiders.[15] But according to the Vietnam veteran, now academic, Andrew Bacevich, FM 3-24 was so vague and self-serving as to be meaningless: "Trafficking in the standard array of postmodern tropes—irony, paradox, bricolage, and sly self-referential jokes, Petraeus's manual [says] next to nothing."[16] Bacevich's analysis was corroborated by the confusion that characterized U.S. military operations in Afghanistan.

That confusion may have eluded Western politicians but it was obvious to Afghani president Hamid Karzai, who, in a damning rebuke of Petraeus and his strategy, stated that the presence of foreign troops in his country "inflamed" local emotions and led "angry young men" to join the insurgency. U.S. raids intended to capture and kill militants, and which necessarily involved home invasions, must "go away," Karzai demanded.[17]

Nor was Petraeus' attempt to recreate in Afghanistan the illusory benefits of the Iraq surge any more successful. In 2010 Petraeus had managed to extract an additional 30,000 "surge" troops from a reluctant president Barack Obama. But when in 2012 the last of those forces were withdrawn, all they left behind was "an uncertain landscape of rising violence and political instability."[18] It seemed no coincidence that the end of this surge was announced by then defense secretary Leon Panetta from the other side of the world, in New Zealand, and that the event was "studiously ignored" by U.S. officials.[19]

Because of its reliance on occupation, mass, and attrition, COIN as applied by the West in Vietnam, Iraq, and Afghanistan has been nothing more than a failed variant of first-generation strategy. That should not be surprising given its premises. The fact is, in today's global village, the theory of COIN is intellectually untenable and operationally unachievable.[20]

Central to contemporary COIN operations is the notion of the three-block war, a concept that attempts to define a model by which invading forces can succeed in an unfamiliar, hostile, often urban environment.[21] It is noteworthy that the model grew out of the persistent failure of Western armies to cope with precisely those conditions during unsuccessful campaigns in Vietnam, Somalia, Iraq, Bosnia, the Gaza Strip, and Lebanon.

The argument is that in any three contiguous urban blocks, soldiers might be required to deliver humanitarian assistance in the first, act as peacekeepers in the second, and fight a life-or-death combat in the third. Having established a foothold in the disputed territory, they are then expected to facilitate nation building through the introduction of democratic institutions, free association, an open press, economic reform, and so on. The model is an accurate enough description of the complex and challenging environment confronted by Western soldiers in the past fifty years. The problem is finding an army capable of satisfying its demands.

The eminent strategic scholar, former Australian army officer, and Oxford professor Robert O'Neill has identified the qualities Western armies require to succeed within this setting.[22] His findings describe a force whose hypothetical standards frankly stretch credibility.

According to O'Neill, successful COIN operations demand soldiers who are able substantially to "erode" the cultural barriers that separate them from the people they are trying to help. In itself that is a sensible objective. But when those barriers are listed as language, religion, social mores, and a knowledge of local history, geography, institutions, and economics, the argument tests belief. And if that were not enough—remembering that in many instances these same soldiers will be, properly enough, in fear of their lives—they also have to master civilian skills (for civic aid programs) and have some capacity to "enter into an informal exchange with indigenes." It is difficult to avoid the conclusion that O'Neill's idealized army is based more on wishful thinking than on an objective analysis of what soldiers can and cannot do.

Former U.S. Marine Corps commandant Gen. James Conway offered a brusque assessment of this model, dismissing it as a "masquerade." Armies are unsuited to long-term nation building, Conway stated, because soldiers are "killers," not "social workers."[23] Conway's pungent commentary on the role of armies should not be taken as criticism of Western soldiers, who remain highly capable when they are doing the job that recruitment standards and training make them competent to do, namely, applying organized violence in the interests of the state. His commentary should, however, be taken as criticism of those senior military officers who, despite fifty years of evidence to the contrary, keep telling us that first-generation thinking will produce twenty-first-century strategies.

Yet still the delusion continues. The latest iteration from the British soldier/scholar Emile Simpson claims to move beyond the model of "bipolar" war between discrete nation-states to a new model of "kaleidoscopic" war between many players, as in Afghanistan, Iraq, and Syria.[24] Simpson makes some sensible observations: for example, he properly locates his own "paradigm" within the domain of "armed political activity," as compared to "armed activity," and he notes the importance within a kaleidoscopic environment of identifying the "true" enemy, against whom one's full attention should be directed.[25] But none of that is new: identifying the true enemy, for example, is simply center of gravity analysis by another name. More to the point, he does not answer the central question; namely, once an occupying army has correctly (we hope) identified whom precisely it should be fighting, how will it go about doing that, given the inherent flaws of first-generation strategy?

The time is long overdue for Western strategists to stop doing the same thing repeatedly and hoping for different results. They could start by seeing the Clausewitzian model for what it is, by understanding what their defense forces are good at and what they are bad at, and by locating their thinking within the world as it is in the twenty-first century.

Fifth-Generation Strategy

In recent years it has become commonplace to identify strike/fighter aircraft by notional generations, a categorization that extends from the first-generation F-86 and MiG-15 of the 1950s to the fifth-generation F-22, F-35, and PAK FA of the twenty-first century.[26] Technological progression through the first four generations tended to be linear, with each cohort generally being faster, better-armed, fitted with improved combat systems, and so on, than its predecessor. There is a belief, however, that the fifth generation is fundamentally different, that it represents a step change over the previous rate of progress. That belief is attributable to the impact of three technologies: low-observability (stealth), which previously was not available; a quantum advance in the fusion of systems that provide knowledge dominance; and a similar advance in the ability precisely to detect, identify, track, and prosecute air and surface targets at substantial distances. In combination, those technologies have radically redefined the meanings of mass, speed, maneuver, and situational awareness.

It is within that context that two of the West's most significant military thinkers since World War II, John Boyd and John Warden, might be regarded as fifth-generation strategists; that is, their work can be seen as representing a step change from Clausewitzian-derived first-generation thinking.

This definitional distinction might be regarded as merely symbolic since, unlike fighter aircraft, strategic thought has no commonly accepted progression through numerical generations. That interpretation, however, would miss the point, which is to emphasize the extent of the difference between failed (first-generation) strategy in the one instance and successful (fifth-generation) strategy in the other.

The lexicon that characterizes successful contemporary strategy is revealing. Fifth-generation strategic writing routinely uses such terms as tempo, the decision-making contest, knowledge dominance, strategic paralysis, the strategic raid, checkmate by operational maneuver, fleeting footprint, effects-based operations, the no-fly zone, and the rapid halt, none of which has any currency in first-generation concepts.

The nature and influence of Boyd's and Warden's strategic thought is the subject of the remainder of this chapter, starting with the distinctively fifth-generation notion of tempo as a strategic quality.[27]

Tempo as a Strategic Quality

In an ideal world, every strategic action would achieve its desired outcome as quickly and efficiently as possible. Precisely how we might measure those descriptors is likely to vary from case to case, but the criteria would usually include minimum costs in terms of casualties and treasure, minimum damage to the environment and to any infrastructure, and an end state acceptable to most parties, including the vanquished. It has been the allure of that kind of rapid, decisive outcome that has motivated such long-standing concepts as the decisive or great battle and the knockout blow and, more recently, the strategic raid.[28]

A central feature of each of these concepts is the priority placed on speed or, more precisely, the rate of application of force. All things being equal, a rapid conclusion offers the best chance of achieving the best outcome, for the obvious reason that the shorter the fight, the less time there is for both sides to kill people and break things. Speed can also make a major contribution to surprise, a tactical advantage which in itself can shorten conflicts.

Speed is a relative term and has a number of dimensions. The Mongol armies that rampaged through Asia and Europe in the thirteenth and fourteenth centuries CE traveled light and drove their horses relentlessly, sometimes covering two hundred kilometers in two days, a previously unheard of rate of advance for armies. Indeed, according to the Venetian explorer Marco Polo, the Mongols could travel up to ten days subsisting only on horse's blood, which they drank from a pierced vein.[29] The ability to appear unexpectedly gave the Mongols a tactical edge and created an aura of unpredictability and terror. Similarly, during the era of sailing ships, it was the unexpectedness of a fleet's arrival on the horizon, perhaps on the far side of the world, that as much as anything made British, Spanish, Portuguese, and Dutch warships the pre-eminent strategic weapons of their age. In other words, for both the Mongols and those naval powers, speed amounted to a de facto form of surprise. While technology has long since made horse-mounted cavalry and sailing ships obsolete, the general point remains valid.

The dimension of speed represented in those examples is that of movement, of being able to get from one place to another within a time frame that, regardless of whether it was two hours, two days, or two years, generated shock. But movement represents only half of the physical aspect of the speed equation. The other half is intensity of force application. An army that creates surprise by arriving quickly but which then is incapable of applying enough pressure for any number of reasons—insufficient firepower, insurmountable defenses, adverse geography, inadequate logistics—may well have wasted the time and effort expended getting from A to B.

Encapsulated partly in the principle of war of concentration, the rapid application of force can, like rapid movement, create shock and overwhelm. At one end of the scale, rapidity of application can be generated by the sheer number of weapons employed, while at the other end it tends to be associated with the continually increasing efficiency derived from new weapons technology. For example, with its rate of fire of five hundred rounds per minute, a single Maxim machine gun from the late nineteenth century was the equivalent of one hundred riflemen; thus, in one engagement during the Matabele War (1893–94), fifty British soldiers equipped with four Maxim guns were able to fight off some five thousand Matabele warriors. As the British man of letters and politician Hilaire Belloc whimsically noted in a more general context: "Whatever happens we have got the Maxim gun and they have not."

An alternative form of rapid application emerged twenty years later in the shape of strike aircraft, with their singular ability to revisit and so reattack the same target in a comparatively short time frame. In the example represented by the Maxim gun and aircraft, the de facto speed of application represented by mass was replaced by the true speed of the weapons system, with its inherent economy of effort.

Rapid maneuver and force application are not ends in themselves; rather they are operational methods, whose purpose is to seize the initiative and throw the enemy off balance, to prevent the enemy from implementing its preferred plan of action while enabling you to implement your own. That is, they are tools to be used in the broader contest of decision making on the battlefield.

There is nothing new in any of this: the competition of intellect and will has always been the essence of applying strategy, and successful strategists have always looked for better tools to serve their cause. What is new is the pace at which that competition can now proceed. And this has introduced a third dimension to the concept of speed. Advanced communications, computing, and intelligence systems have dramatically increased the speed with which information can be collected and assessed, and with which decisions can be made and, ultimately, actions taken, to the extent that tempo has become a strategic quality as fundamental as firepower and maneuver. The difference in the manipulation of tempo in its broader sense is one of the key distinctions between fifth-generation and first-generation thinking, and that difference in turn reveals a great deal about the inadequacies of the dominant (Clausewitzian) Western military mindset.

Experiences from the American Civil War of 1861–65 serve as a starting point here. Two of the more important lessons to emerge from the war between the American states were the enormous advantage inherent in position and the increasing potency of firepower, the corollary of which was that massed assaults were very likely either to fail or to incur terrible casualties, which was precisely what happened. Yet forty years later, during the Russo-Japanese War of 1904–5, those lessons were ignored by Japanese commanders, whose infantry were mowed down by the thousands while making massed frontal assaults against Russian machine-gun posts. It was then the turn of military experts in the West to ignore the Japanese experience by retreating into first-generation thinking. Because the Japanese eventually won that war, the conclusion reached in European academies was not that attacks against such odds and with such immense casualties were no longer acceptable, but rather that they had to be "pressed harder, with more men."[30]

That was the dominant strategic model at the beginning of World War I in 1914, when armies based on massed infantry supported by heavy artillery and machine guns were seen as the key to victory. The consequences were of course appalling.

Toward the end of World War I, maneuver at least began to equal sheer weight of numbers and firepower as the primary desirable characteristic of combat formations. The increasing mechanization of armies was central to this change, as was the emergence of airpower and a quantum improvement in communications systems, which revolutionized a commander's ability to know who was doing what, where, and to whom, not least the commander's own troops. Primitive communications, or more often their complete absence, had been a major factor in the ability of the Mongols and sail-era warships to arrive unexpectedly, to generate speed via ignorance, and therefore to achieve surprise. The implications of the communications revolution for the development of strategy were enormous.

The blitzkrieg technique the Wehrmacht tested during the Spanish Civil War of 1936–39 and then employed to great effect in the early phases of World War II symbolized the return of comprehensive maneuver to the battlefield. In general, fighting in World War II achieved a balance between force application and maneuver, with both being necessary and neither likely to succeed without the other. There were of course exceptions. On the Eastern Front, the Red Army eventually overcame its early tactical ineptitude through sheer brute force based on mass and position; by contrast, operations in North Africa and the Pacific were characterized by frequent movement, often over great distances and sometimes very fast. The maritime nature of the Pacific campaign in particular placed a priority on maneuver, noting that once the Allies had all but severed Japan's shipping lines of communications by mid-1944, the firepower of the emperor's armies that were stranded on bypassed islands became irrelevant to the final outcome.[31] The re-emergence of maneuver was an event of the highest order.

Major wars fought between 1950 and 1990 confirmed this shift. Fighting in Korea initially featured rapid movement and daring maneuver as the competing armies surged up and down the peninsula before settling into a positional stalemate along a front line just north of Seoul in mid-1951 (a standoff that still obtains today). Americans during their war in the jungles of Indochina from 1962 to 1975, Soviets during their invasion of Afghanistan from 1979 to 1989, and Iranians and Iraqis during their clash in the marshlands and deserts of the Middle East from 1980 to 1988, all failed to appreciate the changing relationship between mass, firepower, and maneuver, and they paid a price. In Vietnam, American commanders confused technology and firepower with strategy and were constantly outmaneuvered by their enemy both on the battlefield and politically, an experience subsequently shared by the Soviets in Afghanistan, while the technical and strategic incompetence of the Iranian and Iraqi armies condemned them to a reprise of the static slaughter of World War I.

Campaign planners from all four countries would have benefited from an analysis of the Israeli Defense Force's tactics during the 1967 Six Day War and the 1973 Yom Kippur War. Acutely conscious of their numerical inferiority and geographic vulnerability, Israeli strategists exploited their superior decision making and technology to generate an operational tempo which overwhelmed their Arab enemies. During the Six Day War especially, the speed with which the Israelis collected information, made decisions, maneuvered, and applied firepower was so powerful as to become irresistible. The Israelis made tempo a factor in warfighting the equal of force application and maneuver.

Strategic Paralysis

Drawing on both the good and bad lessons from Vietnam and the Middle East, and incorporating important technological developments in computing, communications, information sensors, aircraft design, and precision weapons into their modeling, in the early 1980s a handful of Western strategists began to develop a strategic concept that implicitly rejected the premises of first-generation thinking. Variously described as hyperwar and control warfare, the concept eventually became known as strategic paralysis. The idea was to subject an opponent to an overwhelmingly swift and precise combination of decision making, maneuver, and firepower, all directed against its centers of gravity. Unable to cope with the tempo and metaphorically outflanked in the clash of wills, the enemy would be strategically paralyzed. That was the theory, anyway.

The concept of strategic paralysis demanded fast, furious, simultaneous precision strikes against an opponent's vital points and focused on breaking the opponent's will from the very start of hostilities. Lesser targets that might lie between attacking forces and the enemy's true centers of gravity would simply be ignored. Speed of decision making, maneuver, and force application were of superseding importance. Also called parallel or concurrent warfare—descriptions chosen to distinguish it from the gradual, incremental, and sequential nature of first-generation warfare—the technique sought the rapid degradation of the whole of the enemy's system, a process that in theory would precipitate strategic collapse. Some analysts believe that prototypes of the model were evident during the predominantly airpower phases of the American-led military campaigns in Iraq (1991 and 2003), Bosnia (1995), Kosovo (1999), and Afghanistan (2001–2).[32]

The conceptual framework for strategic paralysis was provided by two U.S. Air Force colonels and fighter pilots, John Warden and John Boyd. Each developed his ideas as a discrete contribution to strategic thought, although Warden was familiar with Boyd's work, which predated his own. Warden was the architect of the air campaign conducted by Coalition forces against Iraq in the 1991 Gulf War, and Boyd was a flier of legendary skill who developed into an influential strategic thinker. Warden's beliefs were developed exclusively in relation to airpower and focused on the physical element of strategic paralysis and thus were concerned primarily with compellence. Boyd's thinking applied to the art of war generally and focused on the psychological element and thus was concerned primarily with coercion. Regardless of the respective author's intentions, both sets of concepts have broad relevance; that is, they can be used to inform strategic thinking in any environmental setting, especially when the best features of each are combined.

FORM AS STRATEGIC PARALYSIS

Three aspects of Warden's work are especially noteworthy.[33] First, he applied a rigor to target classification and selection that too often had been absent in the past. Central to that rigor was his depiction of an enemy's centers of gravity or vital points as a series of five concentric rings, which he symbolically portrayed as a bull's-eye. While each of the rings denoted a worthwhile target set, the innermost represented the highest value, the outermost the least. Starting with the center, representing the highest value targets, Warden's model identified the five critical components of an enemy's system, as follows: leadership; processes ("those facilities or processes without which the state or organization cannot maintain itself"); infrastructure; population, with the emphasis on attacking the national will, not people themselves; and finally the fielded military forces.

According to Warden, all five rings are present in most socioeconomic systems, and if possible, the entire system should be attacked concurrently. If our own force structure limitations prevent the prosecution of this idealized parallel campaign, then the center—the leadership—should invariably be the priority. In practice, however, it might not always be possible physically to reach more than one or two of the outer rings. For example, by about the middle of World War II, the Allies' superiority in the air and at sea meant that German and Japanese strike forces were capable only of reaching the outer rings of the Allies' systems, a shortfall that inhibited their ability to exert strategic pressure.

The logic is clear enough. Victory depends on convincing the competing leadership to capitulate, and because that effect is more likely to be generated by the application of pressure to an inner ring than to an outer ring, the inside targets must be prosecuted first. This method differed in practice, if not in theory, from most previous military strategies, which generally presumed that any campaign would have to start at the outer ring—that is, with the enemy's fielded forces, which almost invariably would be placed between any attacking force and the more critical rings—and work sequentially inwards, towards the ultimate target of the leadership. Warden's reversal of that traditional approach was captured in the slogan "inside-out warfare."

The second notable feature of Warden's concept was its tacit incorporation of the revolutionary effect of emergent technologies, especially precision-guided munitions (PGMs). In World War II, the destruction of a notional target required nine thousand bombs; in Vietnam, improved weapons accuracy had reduced that figure to thirty bombs; in Iraq in 1991, a single PGM could do the job.[34] This was a technological advance of the first order, one that redefined the meaning of mass. Without that revolution in weapons technology Warden's need for irresistible speed of force application and concurrent operations could not have been met; with it, his model became feasible. The immense efficiency gains inherent in PGMs enabled a comparatively small offensive force to prosecute a large number of targets simultaneously, thus generating a disproportionate effect. Technology provided Warden with the tempo his concept demanded.

The final noteworthy feature of Warden's thinking was the emphasis he placed on understanding the opposition's culture, of "getting inside the enemy's head." The United States' bombing campaign against North Vietnam, for example, had often reflected cultural confusion, with American planners imposing first-world values onto a third-world socioeconomic system. Yet there was little point and even less leverage in subjecting an agrarian society to an offensive campaign predicated on demoralizing an industrial society. Consistent with the truism that war is always a clash of wills, Warden understood that correct targeting relies on knowing what the enemy leadership values. Furthermore, placing the leadership at the center of his five-ring targeting model unambiguously reminded campaign planners where their strategic priorities must always lie.

Warden's work was the subject of considerable criticism following the 1991 Gulf War. The most common reproach was that his methodology was one-dimensional, that its single-minded focus on attacking a pre-planned list of targets was mechanistic and linear. As a consequence, the argument continued, the targeting list became an end unto itself, to the extent that it effectively was the strategy. This indeed seemed to be the case eight years later, during NATO's air campaign against the Milosevic regime in the former republic of Yugoslavia in 1999, an operation clearly based in part on an interpretation of Warden's model. But the problem there was more one of application than of theory. When the campaign began to falter, the response of the American army officer in command, Gen. Wesley Clark, was to demand more and more targets to attack, regardless of their relevance to any ends-ways-means construct, let alone to Warden's hierarchy of strategic importance. One observer later described Clark's bombing campaign as little more than a disconnected series of "random acts of violence."[35] It was only after NATO finally began to observe Warden's fundamental principles by concentrating on leadership targets, including the personal assets of Milosevic and his inner circle of supporters ("crony targeting"), that the campaign began to generate the desired effect.

Nevertheless, Warden's preoccupation with the form of strategic paralysis—with how things are done—at the expense of purpose—why things are done—does reflect a theoretical weakness. While his model indicates that war invariably is a combination of physical and psychological competition, in practice Warden's personal belief that the psychological factor is excessively difficult to measure, or even to identify, led him to direct his effort primarily against the enemy's physical nature. But this is an approach that presumes that as the enemy's physical system collapses so too will his capacity to resist, and this is by no means certain. In brief, Warden's model makes insufficient allowance for unintended and unforeseen circumstances and for people's resilience in the face of adversity.

At the same time, Warden brought an admirable clarity to target analysis. Specifically, his unambiguous focus on leadership—on the minds of the enemy elite—was an overdue reminder that war is indeed a clash of wills. In an era when information technology, advanced communications, and smart weapons are continually enhancing the ability of all forms of combat power to achieve high rates of operational tempo, and are increasing the opportunity for commanders generally to pursue a strategic effect from the onset of hostilities, the significance of this is considerable. If we substitute "center of gravity analysis" for "target analysis"—that is, if we translate Warden's thinking from its comparatively narrow air campaign origins to broad strategy—the true merit of his contribution becomes apparent.

Purpose as Strategic Paralysis

The emphasis on purpose lacking in Warden's model can be found in Boyd's self-styled "universal logic of conflict."[36] Whereas Warden sought to compel by imposing change on the enemy—by denying the enemy a role in deciding how and when conflict would be resolved by physically collapsing its strategic environment—Boyd sought to coerce, to convince the enemy that its best option would be to modify its unacceptable behavior. Reflecting his familiarity with Sun Tzu's philosophy, Boyd noted that the key to winning or losing is always the opponent's perceptions.[37] In terms of ways and ends, Boyd was suggesting that the only purpose in attacking, say, a particular command bunker, or bridge, or power station, or dictator's mansion, is the effect that action will have on the true target, namely, the enemy leader's mind.

Boyd is best-known for his OODA loop, an acronym for observation-orientation-decision-action. A fighter pilot in the Korean War, Boyd was intrigued by the F-86 Sabre's alleged kill rate of 10 to 1 over the MiG-15 in air-to-air combat, given that the performance of the two aircraft was similar.[38] He deduced that the Sabre pilots' success was due primarily to their aircraft's hydraulically operated flight controls and adjustable horizontal stabilizer, which allowed the American fighter to transition from one maneuver to another faster than its Soviet counterpart. After further thought, he began to see the implications of this particular example for competitive decision making generally.

Regardless of whether we are flying an F-86, or fighting a one-on-one gunfight, or leading a section of riflemen, or commanding a theater-level campaign, or practicing grand strategy, or anything in between, the OODA cycle represents a universal logic of conflict. Simply put, the protagonist who is the faster to act intellectually and physically is likely to win. The OODA loop explains this competitive process. As the loop indicates, we first need to observe our opponent to assess what it is doing and how. We must then orient ourselves to the prevailing circumstances; that is, we must assess what we know about our opponent and ourselves, including such things as experience, culture, support, geography, alliances, firepower, and the desired or acceptable objectives. After we have observed and oriented, we decide what to do, and after we make a decision, we act. Immediately after we have acted, the OODA process recommences, as we observe our opponent to assess its response to our decision and action, and we reorient ourselves, and so on, until the particular decision/action contest is resolved, ideally in our favor.

Boyd regarded orientation as the most important phase of the process. Whereas poor orientation is likely to lead to bad decisions, informed orientation is likely to produce good decisions and therefore superior actions. From that, it follows that our first responsibility is to understand the strategic environment. Only then, Boyd argued, is originality of thought and action likely to flourish, a necessary condition if we are to exploit nonlinear thinking and asymmetries in our effort to "find and revel in mismatches."

If orientation is the intellectual core of Boyd's theory, then time is the key to its application. Demonstrating elegantly simple logic, Boyd noted that time simply exists, that it does not have to be transported, sustained, or protected, and that everyone has equal access to it. In other words, time is a free good that a skillful decision maker should exploit and a less skillful decision maker is likely to squander. In particular, time will be the ally of the protagonist who is best able to compress the OODA cycle, who can repeat the loop faster and more intelligently, and who can eventually get inside his opponent's decision-making cycle and thus control the clash of wills.

The criticism is sometimes made that Boyd's model is vulnerable to groups such as terrorists and guerrillas who, the argument goes, can control tempo and invalidate the OODA cycle simply by refusing to respond to their opponent's actions. But this reasoning cuts both ways. On the one hand, the criticism misses the essential nature of the OODA process. Because the cycle is continuous, any decision a protagonist might make, including not responding to its opponent's preceding action, should be observed and oriented. It is then up to the active protagonist to decide on a presumably modified course of action, to implement that action, and immediately to begin the process again by observing the response, and so on. On the other hand, if an opponent's responses continue to frustrate the application of sufficient pressure to induce an outcome, then the objective of imposing strategic paralysis would seem problematic.

The theoretical basis of strategic paralysis is best represented by a combination of Warden's form and Boyd's purpose, with the strengths and weaknesses of the two models tending respectively to complement and mitigate each other.[39] Unsurprisingly, tempo emerges as a critical factor. In theory at least, overwhelmed by the speed with which the enemy is being attacked both psychologically (Boyd) and physically (Warden), the subject will collapse into a state of strategic paralysis.

Applying Fifth-Generation Strategy

The American-led campaigns in Iraq in January–February 1991 and March–April 2003 best represent the translation of the theory of strategic paralysis into practice thus far, in notably different ways.[40]

In 1991 the invasion proper of Iraq began with the most focused air campaign in history, as hundreds of aircraft and surface-to-surface missiles attacked scores of carefully selected targets with unprecedented discrimination. Because the center of gravity as represented physically by the dictator Saddam Hussein and his inner circle was hard to find and hit, the leadership target set was undermined indirectly via strikes against its command and control apparatus. Within days Iraq's communications and information systems were barely functional, and Saddam and his lieutenants were experiencing extreme difficulty in speaking directly to their field commanders, let alone knowing what was going on. Concurrent strikes against key components of Warden's other four target rings compounded their shock and confusion. Saddam had been attacked inside-out, and his OODA loop had been expanded almost beyond control. Well before the coalition armies rolled into Iraq some six weeks later, Saddam's ability to compete in the decision-making contest had effectively been paralyzed. It was no coincidence that the land war lasted a mere one hundred hours.

Before turning to the 2003 invasion of Iraq, an important observation must be made. Military victory is not an end in itself. Notwithstanding the remarkably one-sided nature of the 1991 campaign, President George H. W. Bush's inability to define a satisfactory political end state beforehand and to arrive at one afterwards illustrated that truism. Bush's political failure enabled Saddam Hussein to remain in power, and the Iraqi dictator was soon again perceived as a threat to regional stability and American security. Once Al-Qaeda had attacked America on September 11, 2001, and links, however tenuous, had been asserted between that terrorist organization and Saddam, the presence in the White House of the first president Bush's son, George W. Bush, made a second war with Iraq more likely than not.

As a study of strategic paralysis, George W. Bush's 2003 campaign in Iraq provides a contrast to the 1991 experience. The critical distinction arises from the form of combat power—the means—with which the model was pursued. Following the 1991 war, the key characteristics of strategic paralysis of knowledge dominance, sustained high tempo, precision (discrimination) in force application, and a focus from the beginning on leadership targets had been evident in three other campaigns, twice in the Balkans (1995 and 1999) and once in Afghanistan (2001–2). The force application from NATO in 1995 and 1999 and from the American-led coalition in 2001–2 came predominantly from air weapons, as it had in Iraq in 1991. By contrast, in 2003 in Iraq, the centerpiece of the military campaign was a high-speed dash from Basra to Baghdad by a joint air/land force, with the lead role in the fighting falling primarily to the armies. While some components of the strategic paralysis model were still the province of discrete air operations, notably the attempts to decapitate key Iraqi leaders, this time it was the land forces that sought to overwhelm the enemy by dominating decision making on the battlefield and controlling the operational tempo. That they were just as successful as their air force predecessors had been in 1991 indicated a broader relevance for the notion of strategic paralysis than some critics might previously have acknowledged.

But again an important point must be made, again related to the political dimension of strategy. It may be the case that on occasions a military campaign or strategy can be too successful for its own good, particularly, perhaps, in the case of strategic paralysis. The Nazis' invasion of the Soviet Union in June 1941 provided an early illustration of this apparent contradiction, which was later also evident in Iraq in 2003.

As had been the case during the blitzkrieg through Poland, the Low Countries, and France at the beginning of World War II, the Germans initially swept all before them in the USSR. Among other things, the Nazis destroyed much of the infrastructure of the Red Air Force as they overran air bases and factories west of Moscow. The distinctive nature of this destruction was, however, to create an unintended consequence for the Luftwaffe. Because the Red Air Force was crushed so quickly on the ground, many of its most valuable assets—trained pilots—survived.[41] In terms of center of gravity analysis, skilled pilots are a much more important, expensive, long lead time, and perishable resource than bases, factories, and aircraft. Thus, a year or so later, when the Nazi advance had stalled and modern aircraft were pouring out of Soviet factories that had been rebuilt beyond the Germans' reach, those same pilots were able to return to the conflict and eventually assert their dominance. Had the German invasion been protracted and had the Soviet pilots been killed in a war of attrition in the skies, that could not have happened.

That experience was repeated sixty-two years later in Iraq, but in a different form. So rapid and comprehensive was the U.S.-led coalition's victory that when the military campaign ended after only five weeks, the Iraqi army had suffered relatively little physical damage. Consequently, large numbers of Saddam's soldiers were able to slip away and later join an insurgency that within months had turned the initially triumphant coalition forces into a deeply unpopular army of occupation and that had brought Iraq to the brink of civil war. Once again, it seemed, strategic paralysis had been too successful.

It would of course be facile, indeed immoral, to contend that a less efficient military strategy, such as the campaigns of attrition fought in World Wars I and II, which had shattered the enemy's forces by the time victory was achieved, should have been applied in 2003. A swift, low-cost victory is by definition a good victory. The issue for the Nazis in 1941 was that their blitzkrieg against the Soviet Union came desperately close to winning, and if it had succeeded the immediate survival or otherwise of the Red Air Force's pilots would have been immaterial. Once the decision to invade the USSR had been made, and given the extent to which German forces were already committed in other theaters, the gamble on a quick victory in the east was not only justifiable but probably essential.

Nor should any criticism be made of the stunning military campaign fought by the American-led coalition in Iraq in March–April 2003. The problem there was the failure of the administration of President George W. Bush to reconcile the United States' desired political ends (the democratization of Iraq and then the broader Middle East) with the chosen ways (invasion and occupation) and means (military power). In turn, that problem arose because of the Bush administration's ignorance of Iraqi history in particular and Islamic culture in general, not because of any military strategic shortcomings.

Indeed, if the second Bush administration had possessed a more informed understanding of Iraqi society, the nature of the coalition's military victory—that is, the utter dominance of the strategic paralysis campaign—would have facilitated the process of regime change and political reformation. Because the coalition's center of gravity analysis (targeting) was so good and the period of fighting was so short, most of the essential human capital and physical infrastructure needed to reform and rebuild postwar Iraq was still intact.[42] Regrettably, President Bush and his appointee as first administrator of the Coalition Provisional Authority of Iraq, Ambassador Paul Bremer, lacked the intellectual sophistication this challenge demanded, and their inept postwar management precipitated Iraq's spiral into chaos.[43] Strategic paralysis may have set the stage, but like every strategy it remained subject to Clausewitz's maxim that the object of war is "the political view."[44]

Perhaps the final conclusion we should draw here is that the sheer speed with which strategic paralysis can produce results increases the already powerful imperative for decision makers to understand precisely what it is they are seeking to achieve, and what that end state implies.

Adapting Fifth-Generation Strategy

Fifth-generation strategy was applied by powerful air, space, sea, and land forces at an overwhelming operational tempo during the campaigns in Iraq in 1991 and 2003. That is not, however, the only model the strategy's logic can inform. On the contrary, it is applicable to a wide range of operational settings, and it can be implemented at varying levels of intensity. That few fifth-generation models have emerged thus far is a reflection not so much of the strategy's logic, but rather on narrow-mindedness within Western defense academies. Specifically, the failure to learn from the occupations of Vietnam, then Iraq, then Afghanistan, is indicative of an institutional bias toward first-generation thinking.

Notwithstanding that bias, several models that clearly reflect the rationale of fifth-generation strategy have emerged. Two especially worthy of note are checkmate by operational maneuver, developed by Gen. Robert Scales, USA (Ret.), and the halt phase or rapid halt strategy and its subset, the no-fly zone, developed by the United States Air Force.

Drawing on his analyses of the U.S.-led air campaigns in Iraq in 1991 and the former Republic of Yugoslavia in 1995 and 1999, Scales proposed an operational concept for the employment of highly mobile land forces defined by speed, precision, knowledge dominance, and a fleeting footprint.[45] As Scales acknowledged, he wanted armies to replicate the characteristics of advanced airpower.

Scales concluded that because of the dominance of Western airpower, future opponents would be unlikely to fight in mass (which would leave them vulnerable to punishing air strikes), but would instead seek to follow the classic guerrilla tactic of operating in small groups and making high-value, high-publicity, hit-and-run attacks against civilian as much as military targets. Consequently, land forces would need a different mindset and structure from those that characterized first-generation armies. The most useful soldiers under Scales' construct would be those capable of exploiting information derived from airborne intelligence, surveillance, and reconnaissance (ISR), including Airborne Early Warning and Command (AEW&C) aircraft, surveillance unmanned aerial vehicles (UAVs), signal intelligence platforms, and satellites, and of then leveraging the flexibility and precise standoff firepower of strike/fighters, long-range missiles launched from surface platforms, unmanned combat aerial vehicles (UCAVs), gunships, attack helicopters, and loitering weapons.

Against that background, Scales proposed a combined arms methodology in which armies "would not need to occupy key terrain or confront the mass of the enemy directly."[46] Implicit in his concept was the judgment that in many circumstances it would be preferable either to destroy an enemy's assets, or to strike briefly but decisively against one vital point, rather than routinely try to occupy and seize the enemy's territory.

Under Scales' model, doctrinally and technologically advanced land forces would use fast-moving air and surface vehicles to make rapid and unexpected maneuver one of their primary characteristics. They would also work as an integrated whole with air strike forces, with the lead element at any one time being decided by the enemy's disposition. Should the enemy concentrate, it would be identified and attacked with precision weapons launched from air platforms operating at standoff distances. Should the enemy disperse and go to ground, not only would the enemy negate its own ability to concentrate force, but also would leave itself vulnerable to attacks by numerically and qualitatively superior land forces exploiting their rapid maneuver capabilities. Prototypes of this kind of operation were evident on occasions during the American-led campaigns in Afghanistan in 2001–2 and Iraq in 2003.

For example, in the months leading up to the invasion of Iraq in March 2003, a small group of American, British, and Australian special forces won a remarkable victory. Their immediate objective was to ensure that western Iraq was free of Scud missiles, which might have been fired at Jordan and Israel, thus dangerously broadening the impending war. Not only did the allied forces meet that objective but also they effectively controlled about one-third of the Iraqi land mass. According to the then-chairman of the U.S. Joint Chiefs of Staff, Gen. Richard Myers, the key to that extraordinary achievement was the availability of airpower—surveillance, reconnaissance, intelligence, and strike—twenty-four hours a day, seven days a week, which was fully integrated with the action on the ground.[47]

A crucial feature of Scales' model is the brevity of the occupation and warfighting phases, noting that it was only when Western armies overstayed their (strictly limited) period of usefulness and tried to become something they are not that they started to experience serious problems in Iraq and Afghanistan. It is a matter for regret that Scales' concept seems to have received little attention in Western military academies.[48]

A similarly biased approach to strategic thinking was evident during a contentious debate over the USAF concept of halt phase or rapid halt warfare. A general model for theater-level land warfare involves three stages: first, halt the enemy advance; second, build up one's own forces; and third, launch a counteroffensive. The model is favored by armies, whose innate slowness to deploy and reinforce makes stage two critical if they are to play a pre-eminent role in stage three.

During the 1990s strategists from the USAF proposed a variation to this model, with which they planned to leverage the West's greatest military comparative advantage by substituting airpower for land power. Many third-world armies fight very well close-up, where they can use their massed numbers and disregard for casualties to advantage; the point is to deny them that opportunity. Consequently, the phases of the USAF's version of the halt phase were: first, halt the enemy army's advance rapidly (with airpower); second, punish the enemy's warfighting resources rapidly (with airpower); and third, having seized the initiative, choose any one of a number of options of what to do next, such as attack the enemy leadership, build up one's own forces, pursue a diplomatic end state, impose sanctions, do nothing, and so on.

The rapid-halt model requires the same kind of warfighting capabilities as does strategic paralysis, but it seeks to apply them with a different purpose in mind. Gaining control of the tempo of fighting is the key. When an enemy is advancing, your own forces have by definition lost control of the tempo; thus, the purpose of phase one is to reverse that situation with airpower. You can then manipulate the tempo in phases two and three as you choose.

Military forces can only be as good as their ideas. By any measure the rapid halt represented high-quality thinking and unquestionably warranted an open and full debate. Regrettably, when the USAF chief of staff presented the idea to the Joint Chiefs of Staff, it was rejected outright by the Army. In what to a disinterested observer seems to have been an extraordinarily small-minded action, in 2001 the then-chairman of the Joint Chiefs of Staff, Gen. Henry Shelton of the Army, ordered all references to the concept to be "excised" from joint documents.[49]

Perhaps if the rapid halt had been fully debated it would have been exposed as conceptually and operationally deficient; alternatively, perhaps it would have been revealed as an important addition to strategic thinking. But in a sense, any conclusion we might reach regarding the halt phase concept (or any other concept) is not the point. The point is that in the interests of identifying the best possible strategies, ideas should be encouraged and their merits or otherwise fully examined. Censorship can only stifle initiative and inhibit progress.

As well as leveraging the West's greatest military comparative advantage, the rapid halt was intended to ease a long-standing pressure point in U.S. defense policy, namely, the requirement that they be able to prosecute two theater-level wars simultaneously. While that policy forces be an accurate reflection of international threats and interests, its potential demand on the armed forces has long been a concern for American planners. Among other things, the rapid halt could have addressed that concern. By substituting advanced airpower for massed (army) manpower, the concept would have alleviated the two-war policy's unrealistic personnel requirements.

The U.S. Army's ponderous performance over the past quarter-century in deploying to wars in the Middle East, the Balkans, and Afghanistan lends credence to suggestions that its leaders opposed the rapid halt because they feared it would lead to further public exposure of their service's disturbing inability to move rapidly and to deal with twenty-first century threats.

A measure of poetic justice was dispensed, therefore, when in July 2005 a Pentagon policy review incidentally revived the notion of the rapid halt. Responding to the extreme pressures being placed on military personnel by the so-called war on terror, senior Pentagon planners reportedly began to question the two-wars construct, looking to replace it with a "one conventional campaign, one war on terror" policy as a means of reducing the demand for people.[50] While the rapid halt was not explicitly cited, its relevance to that setting was obvious.

Indeed, only several years after General Shelton had censored the term from the military lexicon, U.S. air strike crews had executed a de facto application of the rapid halt strategy to achieve a stunning victory over two divisions of Saddam Hussein's elite Republican Guard. Early during the second invasion of Iraq, in late March 2003, those divisions were instructed to deploy eighty kilometers south of Baghdad to stop the advance of allied ground forces. Instead, the guards were themselves in effect halted by a concentrated and precise aerial bombing offensive, which ignored a seemingly impenetrable sandstorm to inflict heavy losses.[51]

Another highly successful adaptation of the halt phase strategy has emerged in the past two decades in the form of a series of no-fly zones enforced by advanced airpower in a variety of settings. Nominally imposed to control who flies what, where, and when in specified airspace, a no-fly zone can in fact be much more than that.

Such is the dominance of Western airpower through ISR, knowledge superiority, control of the air, and strike, a no-fly zone can be extended to become a "no-move" and "no-massing" zone by preventing the surface movement of an enemy's trucks, armor, and artillery, and by stopping its soldiers from forming up in strength. In practice, the enforcement of a no-fly/no-move zone represents the first phase—stop the enemy advance—of the rapid halt strategy. Operations Southern Watch in Iraq and Unified Protector in Libya illustrate the point.

Operation Southern Watch was a remarkable campaign that ran for more than ten years, starting in August 1992 when Saddam Hussein breached United Nations resolutions, and continuing until the second invasion of Iraq by an American-led coalition in March 2003. Southern Watch involved the successful enforcement of a no-fly zone for all Iraqi aircraft in the area south of latitude 33° north, a vast space that was dominated by NATO air forces flying primarily from Turkey and off carriers in the Persian Gulf. A similar prohibition was enforced north of latitude 36° north. There is no doubt that, if necessary, the air blockade could have been extended to include vehicular surface movement. In effect, an "expeditionary" force operating from its home bases occupied about one-half of Iraq.

An equally remarkable outcome was achieved by a multistate coalition of air forces during the Libyan civil war from March to October 2011. Ostensibly the enforcement of a no-fly zone, the operation was in fact a form of the halt phase strategy.

According to the UK's then-defense minister Liam Fox, when the UN-endorsed no-fly zone came into effect on March 19, Libya was only forty-eight hours away from a humanitarian disaster. The well-equipped, professional army of the dictator

Muammar Gaddafi had raced across the country from Tripoli to the outskirts of Benghazi, the center of a popular uprising. British prime minister David Cameron feared a massacre similar to the one in Srebrenica in 1995, when eight thousand Muslim civilians were slaughtered. Instead, air and missile strikes stopped Gaddafi's army in its tracks; that is, a rapid halt was effected. Within hours it became untenable for Gaddafi's soldiers to mass, because they were detected by airborne ISR and then decimated by airborne precision strikes. Similarly, the dictator's tanks, artillery, trucks, etc., became targets whenever they broke cover.

Once it became apparent that NATO would hold its nerve and maintain the no-fly/no-move/no-mass zone until the job was done, and that the anti-Gaddafi National Transitional Council's irregular fighters had the courage of their convictions, the rebels' ultimate victory was almost assured. It is noteworthy, incidentally, that while the NTC irregulars received advice from NATO special force soldiers, they were largely untrained and underequipped. The difference, of course, was that they had airpower and Gaddafi's professional army did not. It is also noteworthy that Prime Minister Cameron was told by a range of military experts that Libya could not be saved by an air campaign, but he "stuck to his guns."

It is important to acknowledge that ultimately revolutionary soldiers from the NTC had to capture urban centers through close-up, street-by-street, house-by-house combat. That the Libyan people won the final victory was as it should be. If any one message has been spelled out loud and clear from the West's disastrous occupations of Vietnam, Iraq, and Afghanistan, it is that political and social change cannot be engineered by occupying (foreign) armies—it must be brought about in its own good time by the local population.

At the same time, the conflict demonstrated once again that a motivated force of irregulars can defeat a well-armed professional army as long as the irregulars are backed by the keys to modern warfare, namely, control of the air, airborne ISR, and precision strike. As the veteran Middle East correspondent Patrick Cockburn reported, "The overthrow of Muammur Gaddafi . . . was mostly brought about by Nato air strikes."[52]

The point bears repeating: Military forces can only be as good as their ideas. Drawing on the characteristics of fifth-generation strategy, the concepts of checkmate by operational maneuver and the rapid halt represented intelligent responses to the repeated failure of first-generation thinking. That both have been largely ignored by the broader strategic studies community, and, indeed, that one was officially censored by the U.S. military, is unconscionable.

Commanding Fifth-Generation Strategy

The merits of checkmate by operational maneuver, or the rapid halt, or any other kind of strategic concept will generally be debatable. What should not be debatable is the importance of constantly putting forward and challenging ideas. It is clear from the repeated failure of first-generation strategy for more than half a century that the shameful decision made by the chairman of the U.S. Joint Chiefs of Staff to censor debate on alternatives was inimical to the institutional good. And that is why the subject of "command" is so important to the development of fifth-generation strategy.

For most of the history of human conflict, soldiers have dominated the higher command of defense forces. Closing with the enemy and occupying its territory almost invariably were necessary conditions for achieving military objectives; consequently, the theory and practice of warfare has reasonably concentrated on armies. However, during the past seven decades it has become increasingly evident that war is now concerned more with acceptable political outcomes than with seizing and holding ground, just as it has become evident that airpower has constantly expanded its ability to influence, even control, events in all three environments. These developments imply a fundamental shift in how conflicts should be planned and fought, an implication that was translated into practice during phases of Operations Desert Storm in 1991, Deliberate Force in 1995, Allied Force in 1999, Enduring Freedom in 2001–2, Southern Watch from 1992 to 2003, Iraqi Freedom in 2003, and Libya in 2011. Yet notwithstanding the military success of those phases, the manner in which many of them were commanded (and therefore controlled) is disturbing.

Even though most of the force was applied by aerospace power, the operations were unduly influenced by army generals, often for the worse.[53] Careful reading of authoritative accounts of the campaigns indicates that the commanders concerned had a limited, perhaps even inadequate, understanding of how to plan and conduct a predominantly aerospace campaign. The examples used here to illustrate the point are Gen. Colin Powell and Gen. Norman Schwarzkopf in the first Gulf War of 1991, Gen. Wesley Clark in the former Republic of Yugoslavia in 1999, and Gen. Tommy Franks in Afghanistan in 2001 and Iraq in 2003.[54]

Powell and Schwarzkopf never fully appreciated the strategic nature of the air campaign constructed for them by the USAF and were always preoccupied with the ground phase of the war. Clark's air campaign (which he insisted on controlling personally down to the most detailed level despite his unfamiliarity with almost every aspect of air operations) has been described as little more than a disconnected series of "random acts of violence," in which his response to the desultory results of the early weeks was to demand more and more targets to attack, with little regard to the effect, if any, their prosecution might have; while Franks' involvement in the ill-conceived Operation Anaconda in Afghanistan in March 2002 (which he later described as "an unqualified and complete success," in contrast to the British Royal Marines' judgment that it was "a military disaster") says more about his army background than his joint force command position.[55] Indeed, within some sections of the U.S. defense organization, Franks was regarded as strategically illiterate.[56]

The performance of all four stands in sharp contrast to the mastery of his brief demonstrated by Gen. Michael Ryan, USAF, in Serbia in 1995, one of the few occasions on which an airman has held a significant joint operational command.[57]

As long as airmen are largely unsuccessful in reaching the most senior joint command positions, which historically has been the case, the replacement of first-generation thinking with fifth-generation thinking will be difficult to achieve.

Culture matters.[58] Many of the major conflicts of the past century have had cultural origins: the Spanish Civil War (ideological), World War II in Europe (ideological), the Korean War (ideological), the Arab-Israeli wars (religious), the American war in Vietnam (ideological), the Soviet invasion of Afghanistan (ideological), the Balkans wars (religious), the war on terror (religious), the civil wars in Libya, Egypt, and Syria (religious), and so on. While we can confidently expect that air forces and armies from liberal democracies will not come to blows with each other over their strategic preferences, we should not underestimate the depth of feeling that underpins their respective cultures.

This chapter has argued that armies have too much influence over the development of strategy, the corollary of which is that air forces have too little. Command and control under army terms is a straightforward affair. Army commands (owns) everything, and the main contribution required of air is close attack. This kind of arrangement is detrimental to the development of innovative ideas and strategies and is at odds with the model of warfighting applied with such overwhelming success by advanced defense forces in recent decades. The fact that the essential capabilities of control of the air, precision strike, and airborne ISR have been conceived, promoted, and sustained by air strategists, often in the face of indifference, even hostility, from the other services, should serve as a salutary warning here.

The first chief of staff of the USAF, Gen. Carl A. Spaatz, once dryly observed that while few air force generals think they know how to run an army, most army generals think they know how to run an air force. Yet it is a measure of the continually increasing authority of airpower in all domains that recent air campaigns may have succeeded despite the command of army generals, not because of it.

Since 1914, air strategists have had to fight a constant war of ideas to ensure that their distinctive capabilities have been understood. Periodic lapses in the necessary intellectual effort almost invariably have been accompanied by increased ignorance of, and therefore resistance to, those ideas. If air strategists are to meet their professional obligations and ensure that the battle-winning capabilities they bring to the joint arena are properly represented, they cannot afford to become uncompetitive in the never-ending contest of ideas. Airpower thinking must be broadly based and topical, and it must be widely understood. The key is for air strategists—including those in armies and navies—to foster institutional attitudes that encourage open and full debate and which reward intellectual initiative.

Discussions about organizational methods, coordination practices, staff responsibilities, and so on, make a necessary contribution to the command and control debate, but they do not come near to approaching the heart of the matter. Any half-competent officer should be capable of designing a workable system for conducting those management-related processes. The real issue lies within the constant clash of ideas, because it is those ideas that shape the strategies, concepts of operations, and force structures that determine how a joint force fights.

Turning to military culture, two factors have historically impeded the degree of change implicit in fifth-generation strategy. First, defense forces have tended to develop through evolution, not revolution. Genuine technological breakthroughs that in turn enable strategic breakthroughs have been few and far between, and even when they have occurred they have not always been recognized or they might have been used inefficiently. For example, in 1940, the British and French armies grossly misused their tanks because they insisted on absorbing them into the infantry and cavalry, instead of creating new tank-centered formations, as the Germans had done. Similarly, although it is now a century since military airpower began to change the face of battle, its transformative nature is still not fully appreciated.

And second, the military mindset is notoriously hard to change. The aphorism that generals prepare for the next war by studying the last one says more about military tradition and single-service cultures than it does about the intellectual capacity of military commanders.

Which leads to an issue that within many defense forces seems almost intractable, namely, the limits imposed on innovative thinking by single-service cultures and attitudes.

A revealing illustration of this emotional and organizational barrier to fostering a holistic and relevant strategic philosophy can be seen in the missions the single services tend to define for themselves. Armies, for example, almost invariably list their mission as being simply to win the land battle. While winning land battles historically has indeed been the main activity of armies, in itself it need not represent a desired effect, nor does it define the only significant effect we might reasonably expect an advanced land force to deliver. Thus armies have asserted sea denial (Turkish gun batteries dominating the Dardanelles in March 1915); they have won control of the air (Allied troops capturing Luftwaffe airfields in France following the D-day landings in 1944; Ariel Sharon's armored columns smashing through the Egyptians' ground-based air defense system along the Suez Canal in 1973); and so on. There are so many other cases that the point should be self-evident, but it is so important that it does need to be emphatically made. Navies and air forces similarly tend to couch their missions in strictly environmental terms.

The attitudes this kind of thinking represents constitute a formidable barrier to intellectual progress within defense forces. To extend the line of reasoning, the predilection of armies, navies, and air forces to define their capabilities in terms of their particular hardware such as tanks, trucks, frigates, and fast jets, instead of by effects, entirely ignores the often battle-winning roles played by capabilities that notionally belong within an ostensibly different environmental or warfighting model. The basic problem is single-service tribalism.

There are good reasons why the evolution of defense forces has traditionally taken place within the distinct environments of land, sea, and air. Even now when the influence of information operations and the capacity to act with speed and precision are becoming more evenly balanced across armies, navies, and air forces, there are still well-founded specialist and cultural arguments in favor of the long-standing organizational arrangement. Some forty years down the track, Canada's ill-considered decision to peremptorily combine its three services (subsequently rescinded) is still used by guardians of the old order to "prove" the danger of ignoring history.

It is unquestionably the case that the social compact within a professional, all-volunteer defense force is unique, and that peoples' readiness to risk their lives can be related to their identification with their service and unit, as well as to their commitment to their comrades. Nevertheless, as J. F. C. Fuller once noted, the fighting power of a defense force lies in the first instance in its organization.[59] It would be a mistake of the first order if tradition alone were allowed to stand in the way of intellectual reform. For example, we might conclude that if we started today with the proverbial clean sheet of paper to shape a defense force for the twenty-first century, we would not end up with an army, a navy, and an air force as we now know them.

We would certainly conclude that if we started with a clean sheet of paper to develop a winning strategy for the twenty-first century, we would not end up with a concept based on the unsustainable proposition of invasion and occupation, and involving mass, attrition, a limited ability to maneuver, limited situational awareness, and the centrality of a "bloody and decisive clash of arms." On the contrary, those characteristics are the antithesis of the fifth-generation strategy that has underpinned the West's most productive military campaigns of the past twenty-five years. And it must be the success of fifth-generation strategies rather than the failure of first-generation dogma that informs the thinking of our military commanders.

Conclusion

According to the pioneering scholar of group dynamics and organizational development, the German-American psychologist Kurt Lewin, there is nothing so practical as a good theory.[60]

Regrettably for Western societies, the theory that has dominated strategic thinking in their military establishments since World War II has been entirely impractical. Founded on the nineteenth-century dogma of mass, invasion, occupation, and seizing and holding ground, that first-generation theory relates to a world order that no longer exists. Indeed, in its most recent manifestation as a "modern application" of counterinsurgency warfare, first-generation thinking has been not so much a theory as a cult. It should scarcely be surprising, then, that attempts to translate the theory into practice have failed repeatedly and have led to military, political, and social disasters for all concerned in Vietnam, Iraq (twice), and Afghanistan. In short, the model is broken.

At the same time, a series of campaigns founded on knowledge dominance, tempo, precision, and a fleeting footprint have been extraordinarily successful. That is, the theory underpinning those campaigns has been eminently practical. Expressed through operational concepts such as strategic paralysis, the strategic raid, the rapid halt, the no-fly zone, and checkmate by operational maneuver, fifth-generation strategy has been demonstrably relevant to the world as it is in the twenty-first century.

But, again regrettably for Western societies, the strategy's logic is neither properly supported within defense establishments nor well-understood within the broader community.

The world remains an uncertain place, and there will continue to be a need for strong and effective defense forces. Fifth-generation strategy leverages the West's greatest military comparative advantage, and it works. It says a good deal about the dominant defense mindset that opposition to the model has been internal not external, and institutional not intellectual. If Western strategic thinking is to progress, experience indicates it is time for a different military culture and different leaders.

CHAPTER FIVE

Airpower Theory

Colin S. Gray

Airpower theory has suffered from two persisting lethal defects: it has been both logically unsound and empirically fragile, or worse.[1] The second of these enduring problems has been more than marginally the result of the first. Confusing and unintentionally misleading language and argument understandably have had profound difficulty explaining events and behavior. Although there will always be scope for differing judgments on specific issues, the meaning of airpower yesterday, today, and tomorrow is neither mysterious nor should it be particularly controversial. The century-plus of airpower history to date can tell us all we need to know to understand this kind of power. To borrow and adapt from Antulio J. Echevarria's persuasive judgment on Clausewitz, the strategic narrative of airpower is complete yet unfinished.[2]

Although the character of contemporary airpower is always changing, and every situation wherein airpower is applied is unique, the whole subject can be revealed convincingly in a general theory. The theory of airpower presented here is able to cover the subject completely. The nature of airpower can be explained now, even though the technical and tactical stories are ever-shifting. I am not claiming that this offering is or should be the final take on explaining airpower. That task, which is the role of airpower's general theory, can always be done in different ways, including some that may constitute improvements on what is offered here. The validity of the general theory of airpower—unlike airpower strategies, operations, and tactics—is not hostage to particular technical or other judgments. Whatever the marvels of technology in the years to come, they will not change the nature or the strategic narrative of airpower. At least, they will not do so provided that narrative is explained competently.

With respect to theory itself, a warrior-scholar at the USAF's School of Advanced Air and Space Studies, Harold R. Winton, has made the constructive suggestion that theory should accomplish five basic missions: define the field of study, categorize the field's constituent parts, explain how the parts relate to one another, connect the field of study to other human endeavors, and seek to anticipate how changes in the future may affect the field of study.[3] This is excellent advice. The challenge is to do it for airpower in such a manner that it can serve as a complete, if forever unfinished, general theory that is both broad enough to cover all the relevant phenomena yet sufficiently specific to avoid banality.

Context tends to be sovereign over allegedly inherent capability, which is why it is perilous to draw large lessons from what amounts to strongly exclusive historical evidence. This truth was captured economically for us by John Boyd, with his placing of emphasis on the "orientation" element in his OODA loop. Not only should one resist the temptation to celebrate the particular believed lessons of a recent clash of arms as eternal truths about airpower, also one should not do so for the apparent evidence that can be drawn convincingly from a whole decade or two of airpower experience. This is one reason the theory presented below takes the entire period of heavier-than-air flight as its historical domain. The other reason is that I believe airpower should be regarded most usefully as having a single nature whose character is ever-changing and changeable. Of course, airpower theory is written today in a distinctively contemporary form. And all theorists of airpower, no matter how sincere their intent to be general in probing the subject and encompassing all relevant historical experience, cannot avoid being uniquely acculturated persons with attitudes and opinions fairly specific in time, place, and concerns to their own historical context.

Even when an explanation for understanding is pitched at the exalted level of a general theory, claimed to be universally and eternally true, it is advisable to be less than fully reverential. What follows is not a "credo"; it is only an explanation of airpower to provide necessary understanding. Airpower theory is not religion/faith, philosophy, doctrine (except with regard to its value for education), legend, or myth.[4] But it is a serious attempt to provide sufficiently reliable (i.e., in most cases) truth that is admittedly only social (soft) science resting upon history and logic.

In the interests of clarity, accuracy, and practical utility, airpower theory is presented here in the form of many dicta. Each item is a dictum, meaning simply a formal pronouncement. A *dictum* is a statement that is considered, seriously evidenced, and even claimed as authoritative; the term is chosen because it carries less baggage than does *principle* or *law.* A dictum is much more serious than an opinion, but it carries neither the weight of a principle nor the asserted true authority of a law. Each dictum is distinctive, but many overlap with others. This is deliberate and even desirable, because the overriding purposes of the exercise, as noted already, are clarity, accuracy, and utility, and for those it is essential that connections be specified. Many potentially combinable dicta are separated here so that significant nuance is not sacrificed. The peril is the risk of an economical PowerPoint-able airpower theory that would be tersely and elegantly simplistic, and therefore wrong.

The theory here is as complete as I can usefully make it. But it is work that can never be finished, and, for certain, the dicta chosen can be amended by addition, deletion, or combination ad infinitum. And lest there was too much modesty above, it should be understood that although the dicta are not claimed to be principles or laws, they are advanced as statements that should be regarded and treated as permanently valid.

The theory of airpower is presented in the form of twenty-seven dicta. Each dictum (D) is stated as tersely as is compatible with clarity and utility and is augmented by the necessary explanation and illustration. The list immediately below provides a convenient check on dicta subjects.

Airpower Theory Dicta Subjects

1. The general theory of strategy and airpower theory
2. Education for practice
3. Theory and doctrine
4. Definition
5. Aircraft, air forces, and the Air Force
6. Dedicated Air Force
7. Warfare, geography, jointness
8. Attributes
9. Strengths and limitations
10. Strategic value
11. Control of the air
12. Strategic commons
13. Control of air, land, and sea contrasted
14. Unity of air and airpower
15. Strategic effect
16. Strategic value
17. Airpower supporting and supported
18. Strategic and operational perspectives
19. Offensive and defensive instrument
20. Single historical narrative, single theory
21. Targeting
22. A revolutionary instrument
23. Parallel operations
24. Aerial bombardment
25. Technology
26. Space power and cyber power
27. Air forces differ

D1: Airpower theory is subordinate to the general theory of strategy

Airpower theory is not an alternative to general strategic theory. The latter is not discretionary; it is always authoritative. No matter how revolutionary airpower is or appears to be in its nature, character, and consequences, it has not, will not, and indeed cannot revolutionize the nature of strategy, war, or statecraft. But it certainly can and plainly has led to changes in the character of warfare that warrant the adjective *revolutionary.* The general theory of airpower is a specific application of strategy's general theory to the air environment. The technological, tactical, and operational details and the strategies of warfare must vary among the distinctive geographical domains, but there is only a single template for strategy; conceptually it must organize every domain.

D2: Airpower theory helps educate airpower strategists; it is theory for practice

Just as the general theory of strategy has as its primary function the education of strategists who are charged with devising actual, historically unique strategies, so airpower theory serves above all else to educate air strategists so that they are able to meet their distinctive challenges competently. Fighting power most essentially is the compounded product of three principal elements: material, intellectual/conceptual, and moral. Airpower theory alone cannot deliver superior airpower, but it can help ensure that the air agent of policy is employed in ways that are strategically intelligent. Though general in nature, airpower theory often points the way to the kinds of solutions that may work in addressing the practical challenges of the day.

D3: Airpower theory educates those who write airpower doctrine and serves as a filter against dangerous viruses

Theory is not doctrine. The purpose of theory is to educate, while that of doctrine is to instruct in a more or less mandatory way. Doctrine, meaning that which is taught authoritatively as believed best current practice, is an ever-moving story, and the pressures of the present can be so insistent that it is essential for the doctrine-writing process to be impregnated with the structural perspective of the *longue durée* provided by airpower theory. In reality, airpower theory and airpower doctrine overlap as their focus ascends from the tactical through the operational to the strategic. Because of the key roles played by individuals and their creativity, or lack thereof, and the uniqueness of strategic challenges, consideration of the use of airpower above the tactical level calls for judgment that is creatively strategic. Where that is so, plainly one has left the high utility zone for any meaning of doctrine that leans toward mandatory instruction rather than discretionary guidance as advice.

D4: Airpower is the ability to do something strategically useful in the air

Both parts of the compounded concept of airpower are problematic. This is the main reason why a consensus on an authoritative definition has been hard to attract over the course of airpower's first century. *Air* is problematic because it can be held to include capabilities of many kinds that are able to contest aerial passage (e.g., ground-based air defenses) and machines that fly aerodynamically and otherwise (some kinds of missiles). *Power* is problematic because it can refer to capabilities—both military "teeth" and their ground (space and cyber) support of all varieties of demonstrated or credibly potential performance in action—or to some relative metric (e.g., U.S. versus Chinese airpower).[5] Experience in historical assessment as well as common sense should advise us that the all-too-popular comparative listing of airpower that simply cites aircraft numbers is always likely to mislead and frequently to mislead massively (e.g., the Luftwaffe was very much a shop window air force; it was not permitted to divert resources to the spare-parts inventory that drives serviceability for operational readiness). Any definition of airpower that proceeds in detail much beyond the adapted Billy Mitchell formula favored here as D2 immediately is in peril of passing Clausewitz's conceptual "culminating point of victory."[6]

D5: Airpower is aircraft and air forces, not only the Air Force

There is an unanswerable case for an organizationally and legally independent Air Force (see D6), but this fact does not detract from the authority of D5. In the logic, if not quite the letter or the spirit, of Mitchell again, it should be obvious that the nature of airpower is not directive over issues pertaining to color of uniform. Of course, there are highly important questions with immense practical military, strategic, and political significance relating to a security community's choices for airpower ownership and control. For just one example, I believe that World War II most probably would have been lost had the Royal Air Force not been independent of the British army. It is close to inconceivable that the RAF Fighter Command of 1940 could have been created and directed (air generalship) effectively had it, or its hypothetical equivalent, been a part of the army.[7] It is however a general truth that a country's airpower should be understood inclusively, not exclusively, and it should be assayed for everything strategically useful that flies. Often in historical practice it has mattered significantly which color uniform is in the cockpit, but that sometimes regrettable reality cannot negate the importance of appreciating airpower as inclusively as possible. There will always be some grounds to argue about such academic issues as when is an aircraft an aircraft rather than something else and exactly where does the aerial domain meet earth-orbital space?

D6: Airpower requires a dedicated air force, though not all airpower needs to be air force

Although there is an essential unity to a community's security problems and certainly to any episode of warfare, there is both a distinctive grammar to the preparation for and conduct of warfare in the air and a unique strategic perspective derived from an aerial focus. Similar judgments pertain to the land, sea, space, and cyber domains.[8] In some of its roles airpower rightly is to be regarded as forms of land power or sea power that happen to fly. But the authority of that judgment is limited by implications of the geophysical unity of the air domain as well as by potential opportunity costs. Airpower can be and sometimes needs to be flying firepower, ambulances, and trucks functioning as an integral component in the land power combined-arms team. Moreover, it is absolutely necessary that the requirements for direct and indirect airpower support for land power and sea power be provided in ways and with means that reflect the respective realities and needs of land and sea warfare. That said, in addition and sometimes even instead, there can be an advantageous principally air-oriented character to warfare as a whole, as well as specifically in its aerial tactical and operational detail, that soldiers and sailors are not likely to identify or grasp fully. Broad national security problems, as well as particular challenges, need to be addressed by defense professionals educated in the nature of airpower and its contemporary character. Airpower as land power and airpower as sea power in practice are not adequate as intellectual centers of gravity for determining how best to develop and employ airpower. The unarguable and therefore banal fact that all conflict ultimately must have terrestrial reference because man can live only upon the land simply loses the strategic plot. It invites the strategist's most classic question: "So what?"[9]

D7: Warfare is joint, but physical geography is not: The air domain is different

There can be no reasonable dispute over the necessity for military jointness and even the strong desirability of some integration beyond jointness. Nonetheless, the logic of warfare cannot command a fully matching logic of geography. There is a different grammar, though not ultimately strategic logic, to military effort in each of the five unique environments. Armies, navies, air forces, space forces, and cyber forces all have the same nature strategically, but they also have thoroughly different natures because of the specific physical conditions in which they must function. Air forces are different from armies in almost all respects save for their ultimate purpose in war, which is to influence the will of the enemy by the strategic effect of their several more or less well-joined and integrated but certainly complementary behaviors. Armed forces necessarily specialized for each geographical domain do different things differently. It is not impossible, but it is improbable, that a lifetime of professional focus upon warfare on land or at sea will prepare a person as competently to understand how airpower can be employed most effectively as would a lifelong commitment to airpower.

It is well to recall the old advice against correcting so well for one kind of error that one promotes a yet worse kind. To explain, every armed service dedicated to a single geography, or at least not confused about its primary domain of concern, is virtually preprogrammed to err in exaggerating the relative strategic value of its particular domain. Air forces, both independent and especially those still politically aspirant to be so, have been more guilty than the other services of extravagant claims for strategic primacy. This undeniable and regrettable historical fact, however, in no way weakens the argument that it is essential for the aerial dimension to national security to be considered by the people and organizations who by education, training, and experience should understand it best. The plain record of some poor, inadvertently self-harming airpower theory is no justification for a country gratuitously denying itself the net benefits of superior airpower understanding.

D8: Airpower in its very nature has fundamental, enduring, though variable attributes that individually are unique, especially when compounded synergistically for performance

It is commonplace for airpower theorists, and especially for official Air Force doctrinal publications, to itemize the fundamental "core characteristics" of airpower. This is of little value to air professionals who know it intuitively by education, by osmosis, and by experience, but it is vitally important that the non–air professionals, who constitute a substantial majority in the defense community as well as in society at large, be educated as to the nature of airpower. That nature cannot be grasped securely unless its enduring attributes are appreciated. It is probably true that the very familiarity of these attributes tends to work against their being understood fully in their strategic implications. It is important to note that there are two lists of advertised attributes, or claimed core characteristics, on sale in the strategic marketplace of ideas, but only one is to be trusted fully; the other is best disregarded.

The reliable list of airpower characteristics itemizes strictly enduring physical features: speed, reach, height, and as a consequence ubiquity, agility, and concentration. So says Britain's RAF plausibly, indeed unarguably.[10] In its most basic doctrine manual the RAF proceeds honestly to balance its argument by recognizing, admirably, that airpower has some variably important but enduring limitations: impermanent presence, limited payload, fragility, cost, dependence on bases, and some vulnerability to the weather. The second, unreliable list of airpower's alleged attributes shifts fatally between essential geophysical truths and arguable and unnecessary strategic assertions. Specifically, for example, Richard Hallion identifies the following as airpower's attributes: height, reach/range, speed, mobility, payload, precision, flexibility, and lethality, whereas the RAF was safely physical in its argument.[11] Hallion registers claims that, at best, are potentially misleading and certainly are implausible to most non–air professionals. Airpower is not uniquely precise in what it can do, and neither is it characteristically lethal in clear and sharp distinction to land power and sea power, let alone to space and cyber power. Thus does an important and valid argument become infected by a virus of judgment that it does not need. There is no disagreement that contemporary airpower can be precise and deadly, but that does not mean that precision and lethality are unique and eternal attributes of airpower.

D9: Airpower has persisting characteristic strengths and limitations

It is plausible to argue that airpower has enduring, indeed characteristic, strengths and limitations. Moreover, it is compelling to maintain that its strengths have grown much stronger over the course of its century-plus of existence, while its no less characteristic limitations have been addressed so that they have become ever less limiting. I am grateful to Philip Meilinger for that twin-barreled insight, and I find it well-supported by the evidence of historical experience.[12] That granted, the fact remains that airpower does have characteristic strengths and limitations that derive far more from its nature than they do from context. The contexts within which airpower must operate are highly variable, but what follows applies to all cases, albeit with differing potency. A useful way to translate airpower's abstract generic strengths and weaknesses into meaningful strategic ones is to organize a four-way split of categories in answer to the basic question of strategic utility. I choose to ask: What uniquely can airpower do, what can airpower do well, what does airpower tend to do poorly, and what is airpower unable to do?

Opinions are certain to differ over some of the detail, but what follows should command a near consensus. It need hardly be said that some of the enduring features listed, though still authoritative, have been altered almost beyond technical-tactical recognition over the years. Most obviously perhaps in the category "What does airpower tend to do poorly?," state-of-the-art airpower today is vastly more capable of discrimination in targeting than used to be the case.

What uniquely can airpower do? Directly assault physical centers of gravity regardless of their location; attack the enemy inside to outside from his center to his periphery; project force rapidly and globally; observe "over the hill" from altitude (admittedly, this is not unique; it is a capability shared with space power); transport people, modest levels of equipment, and supplies rapidly and globally; and insert and sustain small isolated expeditions, raids, and even garrisons.

What can airpower do well? Protect friendly land and sea forces and other assets from enemy airpower; deter and be the decisive strategic agent for high-level and mid-level regular and conventional conflicts; compensate effectively for (some) deficiencies in friendly land and sea forces; deny or seriously impede enemy access to particular land and sea areas; and deny enemy ability to seize, hold, and exploit objectives.

What does airpower tend to do poorly? "Occupy" to control territory from the air alone; send clear diplomatic messages; close with and grip the enemy continuously; apply heavy and potentially decisive pressure for conclusive strategic effect in (largely) irregular conflicts; discriminate with thorough reliability between friend and foe, guilty and innocent.

What is airpower unable to do? Cost-effectively transport heavy or bulky cargo; seize and hold contested territorial objectives; accept, process, and police an enemy's surrender.

However, this technical-tactical fact is substantially offset by the reality that today the political, legal, social-cultural, and strategic contexts for the use of kinetic airpower typically are far less permissive than in the past. Readers are invited to amend the detail in my four categories or even the categories themselves as they find most persuasive. It is important to note that this four-way split on strategic utility must be employed only within the framework of the whole of airpower theory and the general theory of strategy.[13]

D10: The strategic value of airpower is situational but is rarely zero

The relative strategic worth of airpower varies widely with types of conflict and geographical setting. All warfare is joint, just as all strategy is a part of grand strategy. Airpower is both enabler and enabled, supporter and supported; exactly how much of each depends upon the ever-different contexts for its employment. There is no general truth beyond the sense in the wording chosen for this tenth dictum. It is possible, indeed necessary, to decide the relative weight each geographically specialized military agent should be able to bring to a particular conflict. But usually it will be the case that the outcome was secured by virtue of a total strategic effect that was the product of truly joint team effort. There is no reason in principle why airpower alone cannot deliver decisive strategic success, but the conditions that permit such a victory are rare. Usually the damage and pain inflicted from the air are strategically decisive only because they enable friendly ground power to seize, hold, and exploit. Damage and pain that are not connected credibly to hostile action of a close and personal kind on the ground tend to be easier to bear than is direct terrestrial engagement. Some death from altitude is less conclusive than is physical presence for occupation.

As J. C. Wylie argues, it is "the man on the scene with the gun" who is in control.[14] Nonetheless, Wylie's powerful dictum should not be applied indiscriminately. There are conflicts wherein a state does not wish and may not need to occupy the territory of an enemy who has been coerced successfully. Any general maxim or dictum that claims to capture the relative strategic value of airpower is either fraudulent or simply erroneous. Circumstances determine what airpower can contribute.

D11: Control of the air is the fundamental enabler for all of airpower's many contributions to strategic effect

Control of the air allows friendly airpower to be all that it can be, to achieve all that it can achieve strategically in particular situations. One may choose to distinguish among the imperious Douhetian concept of command of the air, the scarcely less prideful idea of air supremacy, and the ever-popular notion of air superiority. The core idea, with an obvious maritime provenance, is that friendly aircraft can fly when, where, and how they choose while enemy aircraft cannot, at least not reliably.[15] One is not referring necessarily to an absolute and impermeable air blockade of the enemy but rather to a situation wherein the enemy can have almost no confidence in the ability of its surviving aircraft to execute any mission of strategic importance. It always matters to be strategically sovereign in the sky, but just how much it matters must depend upon the character of the conflict and the potency of airpower in the context of events. States that achieve and exploit control of the air tend to win regular conventional wars, particularly when the principal terrestrial referents are relatively open (i.e., not too much mountainous, triple-canopied, or urban terrain).

The critical point is that although airpower can indeed make strategic history, it is always what might be considered a dependent strategic variable, not an independent one. As a general rule there are practical limits to what airpower can achieve strategically, no matter how preclusive its control of the air or how excellent it is technically, tactically, and even operationally. Airpower is a wonderful, multiuse, strategic tool, but it is not the only such tool that grand strategy needs. Occasionally that will not be true, and airpower all but unaided will serve up decisive strategic success. The Berlin airlift of 1948–49 was just such an exception to the rule; NATO's air war against Serbia over Kosovo in 1999 contestably was another; while RAF Fighter Command's strategic victory in 1940, again arguably, registered a conclusive success for (defensive) airpower alone (though it was only a campaign victory, not victory in the war as a whole, no matter that it happened to enable most that followed to achieve a satisfactory outcome). There is never any room to question the importance of controlling the air, but there is usually a great deal of room for questioning just what quantity of favorable strategic effect such control enables and delivers. When airpower professionals insist upon the necessity of achieving control of the air, as they should, they must be ready to answer the strategist's question: "So what" does control mean for us, and for the enemy, in likely strategic consequences?

D12: Superior airpower enables control of vital strategic "commons"

The air and the sea cannot be fortified against undesired intruders, but they can be blockaded meaningfully. Mahan referred to the high seas as the world's "wide commons," then the only physical medium for global communications.[16] Today the global commons have four geographical and virtual geographical domains: sea, air, space, and cyberspace.[17] Air superiority is always only partial, because of the nature of the aircraft instrument. But that granted, an air force able to control who flies reliably to achieve useful strategic effect and who does not is close to an essential enabler of all terrestrial military operations. When hostile airpower is at liberty to fly where, when, and even how it will, readily visible military effort on the earth's surface becomes difficult or impracticable. As airpower has developed technically and tactically, so control of the airspace over the terrestrial battlespace has become ever more likely to enable victory on land and sea. The seas and the air cannot be employed at sovereign discretion if enemies rule the sky. This was not always unarguably so, but since 1940–41 this argument has acquired ever-greater military authority. It is well to remember that even a magnificent and unchallengeable quality of control of the great global commons of the sea and the sky suffices only to enable the dominant airpower to engage terrestrially where it wishes; it does not guarantee strategic success. Most conflicts require military effort on the ground. Control of the air commons is a priceless strategic advantage, but not all advantages can be cashed for a politically meaningful victory. There is a sense in which orbital space and cyberspace qualify even more comprehensively than the air as a global commons and certainly far more completely than the sea. On balance, however, it is wise not to risk conflating air and space, let alone spicing up the already heady brew with cyberspace as well (see D26).

D13: Control of the air is either essential or highly desirable but differs qualitatively from control on the ground

Recognition of the distinctions in meaning among control as in sea control, control on land, and control of the air is not the minor matter it might appear to be, because frequently the question of what airpower might accomplish more or less unaided has been a live strategic issue. To a soldier, the control of territory has a robust meaning, and for it to be claimed credibly, the enemy truly must be *hors de combat*, or at least resting well out of sight. More to the point, control of the ground means control of the relevant people's behavior, which should mean that the war, perhaps the conflict, is over (if only for a while and in its recent character). For air professionals to claim control of their geographical domain need not carry the implication that the terrestrially bound humans that are their enemies will be ready to "cry uncle" and sign the grand surrender document. The successful soldier can look the surviving enemy in the eye and pose the choice between "surrender or die." Airpower by its nature cannot be that up close and personal. This is not a criticism; it is just a fact of strategic geography. When airpower theorists write quite properly about control of the air and, with excellent reason, extol its virtues, they should never forget that they are sharing usage of *control* with other communities. The worldviews of military tribes differ and find expression in distinctive military cultures and subcultures. The same key words may carry significant differences of meaning among the land, sea, air, space, and cyber domains.

D14: The air is one and so is airpower

This belief was among the sounder thoughts of Marshal of the RAF Lord Trenchard. Regarded as a law of strategic nature, this dictum is all too easy to abuse. But when it is regarded as open to some discretion in practical applicability, it contains a large kernel of essential truth. There is a geophysical unity to the sky that is in sharp contrast with the land and stands in some distinction from the condition of the sea(s). In practice as opposed to theory, of course, airpower is not truly a unity, just as some parts of the sky will be more friendly than others. Nonetheless, D14 claims persuasively that for basically geophysical reasons, it is sensible to think about the sky as a single strategic domain, and about airpower, friendly and other, as a unitary force. By its nature airpower is manifested in highly mobile machines that are able to concentrate and disperse rapidly over great distances and therefore with a reach orders of magnitude more rewarding to military direction than land power or sea power. This is not really a matter fit for debate; it is simply a material, tactical, operational, and strategic reality consequential upon geography.

To operate at altitude, aircraft must have the performance qualities that lend them persuasively to unified command. Some truths, no matter how unarguable at core, lend themselves to abuse in unwise practical application. D14 does not claim that all friendly airpower, of whatever character, ought to be commanded and controlled centrally. But this dictum does insist without equivocation that the essential unity and distinctiveness of the aerial domain and the nature of aircraft imply that airpower should be employed in ways that exploit, rather than contradict, its nature. There is a danger that non-air-minded military people will fail to use airpower as it should best be used. Also, there is the risk that airmen will demonstrate an unduly parochial concern for air-specific matters at the cost of neglect of the challenges facing their joint allies on land and sea. The sensible way to interpret this dictum is to say that the geography of air warfare and the necessary nature and evolving character of airpower demand that a centralizing approach to the air should be the default wisdom. Particular cases will demand and require some dispersion and variety in airpower commitment, but those need to be recognized as tolerated exceptions to the rule of unity. Air strategy should be indivisible.

D15: Airpower has strategic effect, but it is not inherently strategic

Some logical and empirical fallacies are so well-entrenched that they are probably beyond reach by reason. Alas, such may well be the case with the long-standing and notably authoritative claims by some airpower theorists that (a) airpower inherently and uniquely is a strategic instrument and/or (b) some airpower is strategic and some is not (i.e., allegedly tactical). With malice toward none, deep respect for some, and empathy for their historical circumstances, still it must be said that airpower theorists and practitioners have misused the concept of strategy for nearly a century. Inadvertently this has been to the detriment of their cause and to the interests that they have striven to advance. To avoid needless confusion, the logic of dictum 15 is the following.

First, strategic effect is the compounded product of all the behavior (military and other) that shapes the course and outcome of a conflict. Second, in its military dimension, the course of a war is shaped and advanced by the net effect of friendly and enemy behavior. Third, again in its military dimension, which is to say in warfare, all behaviors by all military agents and agencies ultimately have some strategic effect. The logic of this argument was expressed thus by Clausewitz: "But in war, as in life generally, all parts of a whole are interconnected and thus the effects produced, however small their cause, must influence all subsequent military operations and modify their final outcome to some degree, however slight. In the same way, every means must influence even the slightest purpose."[18] Fourth and final, it is appropriate to think of war as having tactical, operational, strategic, and political levels, but those "levels" should not be reified inappropriately. All military behavior is tactical in the doing. Approached collectively on a large scale, the tactical has operational meaning, while the large-scale tactical as operational-level behavior is consequential fuel in greater or lesser (a net positive or net negative) amount for the strategic consequences that are expressed in the political outcome.

The view that some airpower is uniquely strategic both shortchanges allegedly nonstrategic and therefore presumably much lesser airpower and relegates land power, sea power, and now space and cyber power to the second, nonstrategic division of military instruments. The fundamental conceptual error is the proposition that because airpower in one form is able to reach and strike an enemy's center(s) of gravity directly, without first defeating its army and navy, it is therefore inherently strategic. The reasoning is not entirely implausible. In theory, at least, it is possible for airpower alone to coerce an enemy's political leadership. But this speculative possibility does not suffice to render airpower strategic. The reason is because the strategic effect is decided by the target, not by the attacking airpower. To have a very great, even strategically decisive effect, does not miraculously change one's nature from the tactical to the strategic. It should be needless to explain that definitions are discretionary and can neither be true nor false. There is a sense in which "strategic" can be whatever I choose so to label. But the tactical/strategic distinction coined by Mitchell was truly a misfortune for his cause. The principal damage to understanding inadvertently caused by the distinction is, ironically, that the strategic badge can hardly help but discourage strategic planning and thought worthy of the title. After all, if some or all of my airpower is by definition (of operational mission and performance characteristics, especially range and therefore reach) inherently strategic, there is little necessity to think beyond what it might do to what might be the consequences of what it does. The tactical, operational, and strategic thus all are compounded, fused, at the price of the neglect of strategy.

D16: All airpower has strategic value in every kind of conflict

Airpower universally is strategically useful. There is an air narrative integral to every conflict, actual or anticipated. The foolish arguments that have persisted over the years with respect to airpower's strategic value relative to land power in particular have tended to obscure the more significant reality of airpower's true pervasiveness. Modern military operations of any character must have a joint, indeed an integral, air dimension. When ground forces are tasked with the heaviest of heavy lifting missions, as currently in Afghanistan, the "boots on the ground" still are enabled to perform their tasks by airpower in its several forms. To illustrate, airpower provides an air bridge for mobility both to and within Afghanistan, provides and supports all C4ISTAR functions, delivers essential medevac services, and both manned and unmanned—but not unpiloted—provides agile precise firepower. Regardless of who is in command and how one might assay the relative weight of air and ground forces, the permanent reality is that all military effort in all warfare today (and tomorrow) contains an indispensable air component. This was an arguable truth from the 1910s, but it has been an axiomatic truth since the 1930s. The challenge is not in recognizing the merit in D16; rather it is to give the sense in this claim its practical due. It is an enduring challenge to plan and conduct operations so that each geographically specialized military instrument contributes all of which it is capable in the specific context of the day.

D17: Airpower both supports and is supported by land power and sea power (and space power and cyber power)

The relationship between land power (perhaps better expressed as ground power) and airpower has shifted over the years, as the latter has become an ever more reliably lethal tactical instrument.[19] A superior, even supreme, airpower was a decisive advantage for the Western Allies in World War II, provided the ground forces were fit to exploit the opportunities that airpower gave them. As those ground forces cashed the tokens earned by airpower, so the latter in turn could provide yet more critically enabling assistance. Airpower enabled ground power to take and hold territory, and the territory thus seized and defended could be exploited to provide hasty airfields, which enabled . . . and so on. It was a cycle of mutually reinforcing success between land and air. To expand the story, sea, land, and air each performed to enable the others. As the title to a famous U.S. Marine Corps memoir put it, there was a "jungle road to Tokyo."[20] But the Marines' jungle road comprised islands connected by maritime and air bridges and seized at the point of combat contact with vital assistance from the airpower of all the armed services.

Recall the significance of the seizure, holding, and use made of Henderson Field on Guadalcanal in 1942, while the controversial assault on Peleliu was prompted by MacArthur's fears about Japanese use of the airfields on the small island. It is appropriate to claim that today superior airpower rules the ground battlespace in open terrain. This means that locatable enemy ground forces should be reliably defeatable by a state-of-the-art airpower that enjoys control of the air. For contemporary airpower to defeat enemy ground forces, it may be necessary for friendly ground power so to menace the enemy that it is obliged to maneuver, mass, and reveal itself, as notably did not happen in Kosovo in 1999. Airpower and ground power render each other more lethal. Air menace can induce enemy ground forces to disperse and hide to the point that they are unable to function effectively against massed friendly ground power. Ground menace, in turn, can be so threatening that enemy ground power has no option other than to concentrate and probably move and thereby expose itself to try and avoid destruction in land battle. Such preparation for effective conduct of ground combat must yield targets that a first-class airpower could hardly fail to destroy or paralyze. One must add the caveats that warfare is complex and that it is truly a duel. Friendly airpower that should offer the potency for favorable strategic decision might be greatly weakened in practice by, for example, enemy ground-based air defenses and political, legal, and social-cultural constraints on targeting.

D18: By its nature airpower encourages operational and strategic perspectives, a fact with mixed consequences for good and ill

The inherent mobility of the aircraft that quintessentially comprise airpower encourages a breadth of view that can be constructive or otherwise. Although the sharp end of war is always absorbingly tactical and horribly personal, it is persuasive to argue, with Wylie, that the worldviews of soldiers, sailors, and air force personnel tend characteristically to differ.[21] Every person fighting or in support contributes cumulatively if minutely to total strategic effect, but it is easier to see the bigger picture, read the whole script, from the vantage point of 20,000 feet unimpaired by terrain than from the ground at eye level or from the bottom of a muddy, leech-infested ditch or even from a ship able to see to the horizon. It can be claimed that airpower theorists should be all but uniquely able to think strategically, such being a gift from their military specialty. Unfortunately, to date the merit in this assertion has been more than offset by its measure of error. The mobility, and hence range, reach, and temporal compression, enabled by airpower ought to encourage a matching width, breadth, and generously contextualized view of the strategic world. The historical record, however, reveals that airpower theorists and practitioners have fallen into the "tacticization" trap.[22] Instead of exploiting their high vantage to adopt a truly strategic big-picture view, they have been seduced into confusing their mobile and wide-ranging instrument with strategy itself. If one is permitted to define *strategic* as meaning long-range and as pertaining to menacing an enemy's nonmilitary center(s) of gravity, then it is understandable that one would assert one's uniquely strategic character. It has to follow that if one authorizes a Strategic Air Command, perhaps more inclusively a Strategic Command, one is saying by unmistakable exclusion that all else among one's forces in some vital way is not strategic. The nonsense in this should be obvious.

D19: Airpower is not inherently an offensive instrument; rather it has both offensive and defensive value

The belief that airpower inherently is an offensive military instrument can be dated with certainty to 1916, when it was pronounced by General Hugh Trenchard of the British Army's Royal Flying Corps (RFC) as though it were a law of nature. If airpower has a First Law, a Prime Directive, then plausibly it has been the maxim that it is inalienably offensive. The problem with this now classical maxim—dictum is too weak a descriptor—is that it is incorrect. What is correct is to argue that airpower is a weapon that has had an ever-increasing potential to be employed offensively for high strategic effect. Once one leaves the comfort zone of a faith-based airpower credo and instead reasons strategically, the fallacy in the classic assertion becomes obvious. Offensive and defensive are determined by intent and situation, not by science and engineering. Any weapon or military support system of any kind can be employed either for offense or for defense, or for both. The key question is definitional; what is the decision rule that discriminates? Do the qualities of the weapon (or other) system provide the answer, or should one look to the purpose for which it is used? It is not air forces that are, or even can be, offensive; rather is it the political choices that direct strategy to find the ways in operational designs and to command tactical behavior to realize those choices as goals.

An air force ultimately is an offensive or defensive instrument according to who owns it and what that owner wishes to do with it. Strategic theory advises that while defense is the stronger form of warfare, offense is the more effective. Because of the vastness of the sky, the mobility and reach of aircraft, and their inherent relative fragility, it is understandable why most theorists and practitioners have endorsed the offensive maxim as a deep and sincere belief. Alas, the plausibility and the frequent success of airpower on the tactical offensive cannot remove the error in the belief and the peril that attends it. When airpower is held to be inherently offensive, it is easy to believe that it should always be employed in a tactically offensive manner; after all, such allegedly is the true grammar commanded by its nature. Whether offensive or defensive (effected from several means, air- and ground-based), the stronger is always a matter of exact historical context, not of general wisdom. Better-resourced air forces have been able to defeat, or at least fight their way through, the various kinds of air defense assets deployed by less-resourced foes. But even that claim needs qualification. RAF Fighter Command won a great defensive air battle in 1940 with only adequate, not overwhelming, resources. Belief that airpower properly used must be used offensively is a dangerous conviction unless one is a careful and competent theorist-practitioner. Trenchard's relentlessly offensive doctrine resulted in high RFC casualties in World War I. Moreover, his doctrinal legacy spurred RAF Fighter Command to celebrate its 1940 victory by indulging in large-scale offensive fighter sweeps over occupied France in 1941, with painful results.

There is a time, place, and occasion for the offensive employment of airpower, but it is a serious error to believe that its agile nature and long reach require its employment for offensive purposes. Operational offense married to tactical defense can be the most lethal of combinations for the enemy. By way of historical illustration, consider an operationally offensive bomber campaign wherein the bomber "stream" or "waves" actually are bait to entice or oblige the enemy air force to do battle and hopefully die. This was the story of 1940 when the Germans employed this operational offense/tactical defense stratagem, and it failed them; also it was the story of the Anglo-American Combined Bomber Offensive of 1943–45 when—eventually—it was successful.

D20: The history of airpower is a single strategic narrative, and a single general theory has authority over all of it, past, present, and future

It can be difficult to appreciate fully the true unity of the airpower experience. It is not hard to discern the continuities that bind together Giulio Douhet with John Boyd and John Warden. But the technical and tactical distance between the world's first purpose-built military aircraft flown by the Wright brothers in 1908 and, say, the F-22 Raptor of today is so large that the airpower story can hardly help but threaten to burst out of conceptual bounds that appear inappropriate.[23] It can seem plausible to argue that changes in quantities mean changes in quality also. In other words, when the technical-tactical details of aircraft performance shift radically, as they have cumulatively over many decades, it is surely reasonable to suppose that the airpower instrument needs a new theory, indeed probably several new theories, to explain it.

This is a fallacy. It is a glorious merit of the general theory of strategy that it can cope with all cases and conditions at all times, and the like claim holds good for airpower. The key point simply is that airpower has an an enduring nature but an ever-variable character. Historically specific strategies for the employment of airpower cannot evade the authority of character, but they need always to be crafted for particular circumstances. Lest the point remain obscure, no matter how technically sophisticated airpower becomes, no matter how tactically effective it is, and no matter how dominant a position it assumes vis-à-vis land power and sea power, it can only ever be a military instrument of strategy in the service of political objectives.

D21: Strategy for airpower is not all about targeting; Douhet was wrong

It is in the nature of ground power, and even sea power, to have strategic effect by acting or loitering with attitude in a neighborhood. Airpower has an improving ability to loiter, as it were, to occupy terrain by overlook from the overhead flank. Nonetheless, airpower has usually contributed to strategic effect not simply by "being there," by hovering with implicit menace, but rather by doing something. And the core of that something for many decades has been the delivery of firepower to the ground. To risk oversimplification, to an air person who naturally believes that his most favored military instrument inherently is an offensive and strategic tool of policy, the world of the enemy is akin to a bombing range or even to a dartboard. It would be a calumny to claim that the elemental equation of targeting = (air) strategy captures anything close to the full strategic historical narrative of airpower. But the belief that air strategy is mainly about targeting has a classical authority in provenance traceable to the great Italian at his clearest and endorsed by generations of air theorists since the 1920s.[24] The problem is not that it is an error to focus on targeting; indeed how could it be? Given that the dropping and firing of ordnance is how some aircraft fight, of course targeting is of high importance.

The problem is not with targeting. Rather it is the error in confusing targeting with its effect(s) and in conflating those effects with the whole narrative of warfare and of war itself. Targeting is important. The challenge, though, is to contextualize targeting from altitude (again, this is orientation in Boyd's OODA loop). A significant cost in falsely equating targeting strategy with the strategy for a war is that the "whole house" of airpower, not only the kinetic, will be shortchanged in appreciation.[25] Airpower may well be judged the decisive enabler of overall victory in a war, but rarely will it be able to deliver that success by conclusive strategic virtue of its own unaided kinetic effort directed by a brilliant, or even just a good enough, targeting strategy. This is not an issue of giving credit and assigning campaign medals. Typically, a country's war effort requires that its airpower be all that it can be, and perhaps then some. An airpower whose equipment, doctrine, and training heavily favor firepower over logistical mobility, intelligence gathering, medical evacuation, and so forth is likely to impose damaging opportunity costs on its political owners.

All airpower is strategic. For a major contemporary example, helicopters, too, "do strategy," whether they are fire platforms or flying troop carriers.

D22: Airpower has revolutionized tactics, operations, and strategies but not the nature of strategy, war, or warfare

Some people have difficulty coping with the existential dualism of the radical, even revolutionary, change that airpower has effected along with the continuities in the subjects that airpower addresses. The nature of strategy, of warfare in the execution of strategy, and of war itself, has not been altered by the birth and maturing of airpower. But having said that, also one must say that airpower, the third dimension of warfare (the fourth is time), has transformed the conduct of war strategically, operationally, and tactically. The point in need of emphasis is that airpower is a revolutionary military instrument that gradually but irresistibly has changed the manner in which war is waged at every level. It is not an exaggeration to talk of the air revolution. As John Boyd and John Warden have emphasized, airpower today can enable a tempo to military actions and strategic effects that paralyzes and disables an enemy's operational designs and strategic intentions.

D23: Airpower is uniquely capable of waging geographically parallel operations of war, but this valuable ability does not necessarily confer decisive strategic advantage

By its nature, airpower (land- or sea-based) can reach and therefore touch enemy assets at any distance. Kinetically at least, this quality is not shared with land or sea forces, with the notable exceptions of the ability of the latter two to project various forms of airpower, such as long-range bombardment with missiles. If one believes that to reach an enemy is to be able to deter it by threat, coerce it through pain, or disable and destroy it by brute force, then kinetic airpower must be the magic potion that cures strategic maladies. Unfortunately perhaps, this is not so. Polities are not always deterrable; they may decline to be coerced, and even when heavily physically damaged, they may elect to soldier on and hope for a change in strategic fortune. That said, airpower is a uniquely agile and flexible military tool able to menace or strike at enemies on, approaching, or far distant from the terrain or sea space of close terrestrial engagement. Airpower often generates more strategic effect by harassing or destroying enemy assets at locations distant from the land battlespace than it does, or could, by close air support or battlefield interdiction. Wisdom in that regard, however, is highly contextual. When, between which belligerents, and where are we discussing? The tactical lethality of airpower in parallel strike operations has advanced radically in recent decades, even recent years. If the technical-tactical achievement of such lethality is indeed the golden key to sufficient strategic effect for victory, then one would be talking about kinetic airpower as the emerging, probably the finished, article. But airpower, no matter how competent in the conduct of parallel operations of war, is not axiomatically able to be strategically decisive in conflicts. Nevertheless, on occasions, a leading airpower is able to craft a feasible strategy for a particular war that depends critically and convincingly on the achievement of lethality from altitude. Then, truly, the natural attributes of airpower, in dominant manifestation, should be the leading military edge that decides who wins or even that wins itself with little assistance from land or sea forces.

D24: Aerial bombardment "works," though not necessarily as the sole military instrument that decides a war's outcome

Since the 1920s, airpower theory has been presented by its leading, certainly its noisiest, authors as having at its core what has been known as "strategic" bombing. To strike at the center(s) of gravity of an enemy far behind the lines on land, and distant from the domain contested by naval power, was held to be the unique, and uniquely strategically effective, contribution of airpower. This item of advocacy and faith came to be equated with the whole of airpower theory. The theory had little time or space for the character of airpower that supported armies and fleets directly or indirectly. Moreover, airpower theory was not exactly eloquent on the merits of potential achievements in performance of duties other than long-range bombardment. On one of his better days, and in less Trenchardian mode, Mitchell could be sensible. After all, he did bequeath us the timeless and persuasive definition of *airpower* as anything useful that flies, strictly "the ability to do something in the air."[26] Of course, if one decrees as doctrine that the only strategically useful duty of airpower is long-range bombardment, then one is in deep trouble with empirical evidence. It was not, is not, and will not be true. But because long-range (also known as "strategic") bombing cannot sensibly be regarded as the sole important mission for airpower, most emphatically it does not follow that it is unimportant or that it is always doomed to fail. If approached intelligently, aerial bombardment is likely to disappoint only if unreasonable expectations are held for its accomplishments. Long-range bombing has been strategically useful in all periods in airpower's short history. What was attempted was not always tactically wise; hence, operational and strategic ambitions were frustrated. The result was that many brave aircrew died as innocent servants of faulty doctrine and ill-conceived operational plans. Nonetheless, long-range aerial bombardment always had some, and on occasions major, positive strategic effect.

Aerial bombing as a threat to deter, to inflict damage and pain to coerce, and to paralyze or destroy "works" to provide strategic advantage.[27] It should not be expected to decide by its own unassisted kinetic effort who will win a conflict. It is understandable in the historical context of the 1920s and 1930s why airpower theorists tended to embrace this heroic claim for the unique strategic value of their instrument, but they were ill-advised to do so. Then there was some excuse derived from institutional political circumstances for their extravagant strategic claims for long-range aerial bombardment, but no like explanation is plausible today. The early excessive strategic ambition was the product largely of immaturity: political, technological and military, and conceptual.

Bombardment is important in warfare, but typically, neither the outcome of warfare nor a whole project of which warfare is the defining part is reducible strategically to the consequences of bombing. This is an enduring fact, not a contestable argument. And the proposition that an ever-improving airpower is closing in on achieving true fusion with strategy and war is simply wrong.

D25: The high relative (to land power) degree of technology dependency inherent in airpower poses characteristic dangers as well as providing characteristic advantages

The technical performance of machines have always been literally vital to airpower to a much greater degree than generally is true for land power or even for sea power. This fundamental physical fact can encourage, and historically has encouraged, an affection for technical performance at the expense of paying due attention to the tactical, operational, strategic, and political purposes of the flying machines.[28] Even at the tactical and operational levels, where technical performance is vital—for example, speed, range, height, rate of climb, payload, "stealthiness," and so forth—there can be a weight of focus on airpower characteristics that tends to absorb creative energy that would be better devoted to combat skills and their sensible exploitation by air generalship.[29] Competence in the cockpit does not necessarily equate to competence in combat. And a technologically impressive air order of battle is likely to be wasted if air strategists as commanders are not ready to perform for the purpose of usefully employing technically excellent air assets. A relatively high technological focus by air forces is inevitable, necessary, and indeed desirable. But the balance is wrong if that focus translates in practice into an air force that bears some resemblance to a costly and exclusive combination flying club and science and engineering society at the expense of what should be the dominant features of a fighting force. It is not inevitable that air forces must err in this way, but the theory of airpower should alert airmen to the inalienable danger. The risks of undue fascination with the material tools are far weightier for air forces than for armies, though not necessarily for navies.

D26: Airpower, space power, and cyber power are strongly complementary, but they are not essentially a unity

It can seem unjust to dedicatedly air-minded people that just as their most favored and even beloved air instrument unquestionably came of age in all senses, it was challenged in the rankings of relative modernity. Ballistic and (unmanned and unpiloted) cruise missiles, atomic and hydrogen weapons, and then orbiting space systems were followed closely by a galloping emergence of computer-based cyber power. Despite being the last word in modernity for futurists in the 1920s and 1930s through the Cold War decades, airpower's former status as the leading material icon of the present and future has been successfully threatened by more recent and therefore yet more modern technologies.[30] Different security communities with fairly distinctive dominant strategic and military cultures regarded the new technologies in some characteristic ways in strategic theory, policy, doctrine, organization, strategies, and tactics. For example, the Soviet Union almost naturally absorbed long-range ballistic missiles into its vast artillery park, while the United States, equally naturally, added long-range, land-based ballistic missiles to the order of battle of its Strategic Air Command.[31] Whereas land power and sea power have had centuries and more to find ways in which to function effectively for joint purposes, airpower, as the newly arrived third partner in the joint party or to the joint marriage, had to muscle in and demonstrate its utility (when it was allowed to leave the bench and play in more than minor roles).

The historically and unstable triad of land power, sea power, and airpower now has become an even more unstable quintet with the arrival of an immature space power and yet more teenage (at best) cyber power. The need for intelligent interdomainal relations plainly is pressing, even though the technological and tactical stories are shifting at a pace that outruns confidence in analysis. One must add that whereas there are literally millennia of experience with land power and sea power in action—singly as well as jointly—and there is slightly more than a century of record for airpower performance, combative space power and cyber power are seriously untried in warfare. The feature of immaturity is particularly important to the question of military organization because the plausible case can be advanced that it is most appropriate for space and cyber assets to be housed bureaucratically under the wing—no irony intended—of pre-existing military structures until they mature and are better understood. That position is reasonable and may even be correct, but on balance most probably it is imprudent. The undeniable and critically serious fact that contemporary land power, sea power, and airpower today are vitally dependent upon complementary space power and cyber power would seem to indicate the wisdom in postponing, even just resisting, the undoubted ambitions of the new Billy Mitchells of space and cyber to own their own military domains. After all, for the dual reasons of technical immaturity and pervasive joint dependencies, surely it is sensible organizationally to fuse space and cyber with land, sea, and air, not to encourage, let alone authorize and thereby celebrate, fission.

The practical logic of extensive historical experience may suggest that the same reasons Trenchard and Mitchell insisted upon a legally and politically separate air service should be judged strategically compelling for space power and cyber power. The core of the logic lies in geophysics. The five geographical domains of warfare are physically different, a fact that commands unique tactics, operations, and strategies (though not strategy, singular). It is only sensible that a country's airpower should be developed and employed by an organization dedicated to its understanding and most skilled in its employment. Of course, the variety of tasks in that employment suggests the merit in permitting land power and sea power to provide airpower highly specialized for their unique needs. Airpower, however, is far too important a national strategic and military asset to be entrusted, in the main, to any organization other than an air service whose principal concern is the health of the country's military air assets. Soldiers and sailors certainly care about airpower, but they care even more about their troops on land and their ships at sea and therefore, reasonably enough, about what airpower can do specifically for them. By extension it is understandable that soldiers, sailors, and especially airmen care very deeply indeed and appropriately about the quality of service that they receive from space power and cyber power. But these are not the right people to be entrusted with ensuring that national space power and cyber power will be all that they could usefully strategically be and become. The fact that space power and cyber power today need to be integral to terrestrial warfare should be no less self-evident than the fact of genuinely joint, and sometimes more, airpower on the national strategic team. An organization dedicated to space power and to cyber power is likely to advance understanding and capability, not least for joint effectiveness, more rapidly than an arrangement whereby space and cyber concerns are not the primary foci of loyalty and concern.

D27: One character of air force(s) does not suit all countries in all circumstances

There is a single theory of airpower, but the shape and size of a particular country's air force are highly variable. Not only does individual context suggest uniquely tailored strategic needs, but also each belligerent airpower is both enabled and constrained in its acquisition of air force by its unique attributes. Bizarrely perhaps, it is enlightening to adapt Karl Marx's famous fundamental socialist principle of distributive justice, "from each according to his abilities, to each according to his needs," for our purpose here. Translated, each polity develops or otherwise obtains the airpower that its unique circumstances require, at least insofar as its assets (or attributes) allow. The point is to insist upon the individuality of a particular country's airpower needs in the context of the enablers and constraints that permit those identified to be realized in practice. Many strategic needs identified by many countries and for much of the time are less than perfectly matched with operationally ready capabilities. The strategy function of ends, ways, and means is logically impeccable, and its recognition is essential for discipline in behavior, but it is an ideal that frequently is not realized in state practice. Because the character of airpower at all times offers some alternative to those buying an air force, there is always fuel for argument over the wisdom in the choices made. What must be appreciated is the base argument advanced in D26 that a state has both a particular set of strategic needs—which will vary over time and in different strategic contexts even at the same time—and a constrained ability to purchase what it decides it needs and then employ it optimally.

If we add a further thought by Marx, the theory of airpower is usefully enriched. Specifically, in "The Eighteenth Brumaire of Louis Napoleon," he advised, "Men make their own history, but they do not make it as they just please; they do not make it under circumstances chosen by themselves; but under circumstances directly encouraged, given and transmitted from the past. The tradition of all the dead generations weighs like a nightmare on the brain of the living."[32]

Countries cannot necessarily buy the airpower that they need in good enough time for when they need it. A careful comparative study of historical cases of airpower failure concluded that such failure was rarely the product of some shortfall in performance "on the day," but rather is more plausibly attributable to long-term structural weaknesses in defense preparation.[33] British air generalship was good enough in 1940; certainly it was better than the Germans'. But the RAF's decisive advantage in that critical year had been earned by virtue of twenty-three years of serious systemic preparation. Polities that seek to improvise effective airpower to meet unanticipated strategic needs are apt to fail. An air force competitive for the demands that may well be made of it always requires years and ideally benefits from decades of preparation. Moreover, it is necessary to remember that an air force fit for the fight and to support the fight, whatever that means in context, is not synonymous with such ingredients of airpower as tax revenue, science and engineering industry, or even military organization and equipment handling skills. To fly well is one thing; to fly well in combat, at least in aid of combat, is something else that requires focus on the dueling aspect of all warfare. The history of air warfare reveals that the best pilots have not always been the most effective fighting ones. Flying skills and killing skills are not synonymous, valuable though the former is for the latter. For a similar thought applied on a larger scale, although airpower can only be grown from much money and advanced science and technology, still it has to be created, sustained, and prudently modernized. It is not sufficient for a polity to be wealthy for it to be secure; indeed wealth may well fuel insecurity. Money, knowledge, and material must be committed years in advance to produce airpower. It is the nature of airpower that it has to be a long-term project.

While both history and logic argue for the importance of an independent air force, it cannot be denied that the USAF and British RAF are by no means the only institutional models that have proven effective.[34] States with defense budgets far smaller than the American or even the British can be attracted to the economies attendant upon assignment of a country's airpower to a single umbrella air force. Alternatively, there are countries that are comfortable subordinating all kinds of airpower to their army and navy. Such control by land and sea power is fundamentally unsound, but it is prudent to recognize the fact of strategic exceptions to the rule of the general necessity for air to be independent or at least considerably autonomous. D27 should be treated as a reinforcement of D6. That dictum in principle endorsed, perhaps condoned, distribution of a state's airpower among several air forces (corps, components, or elements). Empathetically, however, it asserted the virtues of a dedicated air force. What is essential is that a state's airpower should be developed, commanded, and controlled by the military professionals who best comprehend how it can be employed effectively. Institutional forms and command arrangements that weaken the relative influence of air-minded professionals over airpower's quantity, quality, or tasking are inherently undesirable.

Conclusion: Theory Rules!

The evidence of historical experience is in for airpower theory. As strategic history continues to roll, new episodes in unique contexts fuel more debate over many aspects of airpower: Air strategy, airpower in strategy, choices among technologies, airpower generalship and performance, and so forth; there is no end to the stimuli for debate. More than a century of extensive and intensive experience with airpower, however, provides ample—indeed probably redundant—evidence on the basis of which to construct general theory. There is no need to wait on events, anticipate further technological change, or test more hypotheses. The century-plus from 1903, more realistically 1908, to the present can tell all we need to know for us to make sufficient strategic sense of airpower.

In a few major elements the theory of airpower offered here may appear heretical to those who sincerely have been content to adhere to some classic items of faith that I judge, respectfully but firmly, to be unsound. Anyone writing airpower theory today has a great deal of rewriting to do, because some large conceptual weeds have been allowed to prosper in airpower's intellectual garden. Longevity of ideas, even if not especially poor ideas, confers an authority of its own. To adapt what Marx so aptly wrote, the poor ideas of dead generations of airpower theorists can weigh heavily indeed.

Airpower theory should be permitted to educate only in how to approach the actual challenges of ever-changing airpower. The theory ought not to be raided for direct value as added authority in aid of some contestable preference today. If airpower's general theory is deployed to do battle on the issue of the day, it is nearly certain that it will be abused, misused, and, as a result, suffer loss of authority. Airpower theory can guide us only in how to think, not in what to think. It yields words as concepts that have explanatory power about essential matters and how and why they interact. But concepts require application; they should not themselves be commanded to serve in debate over discrete historical strategic choices.

It can be difficult to protect the general theory of airpower from the enthusiastic advocates of emerging, if not already emergent, strategic narratives that allegedly are in the process of overturning the severely limiting real constraints of yesterday. The challenge for the general theorist of airpower is to attempt to fuse change and continuity in a theory that is true enough to satisfy both radicals and conservatives alike. It is my belief that the one hundred-plus years of airpower history have repeatedly revealed a strategic narrative that allows for explanation by means of a single grand theory. I claim that the general theory expressed here in twenty-seven dicta is effectively complete, even though it should be regarded as unfinished in all advisable detail. I am not unmindful of the likelihood that some theorists will believe quite sincerely that their particular beliefs concerning the present and anticipatable future condition of airpower require some radical rewriting of airpower theory. I am empathetic to such a desire, indeed demand. The theory presented above, however, rests on a century-plus of empirical historical evidence and, as importantly, on an understanding of the enduring nature of strategy over time. Periodic arguments in favor of recognizing radical changes in political and strategic behavior not untypically have considerable merit. But it remains the case that the human political and strategic estate continues to be obedient to the general theory readily accessible in the pages of Thucydides, Sun Tzu, and Clausewitz. Of course history has marched on in the great stream of time since their eras, but arguments for true revolution in strategic theory that is prepared to dump key elements in these classics invariably overreaches.

Airpower has demonstrated the ability to achieve revolutionary change in the character of warfare, of that there can be no reasonable doubt. But it has not, and is not likely to, alter more than warfare's character. Radical theorists always are prone to believe that they can break the code of long-settled thought and practice, which is admirable and desirable. But there is a general theory explaining the nature of politics and of strategy, in the context of continuity in human nature, that always is constraining of the practical possibility of change. For recent examples in illustration of this key logical point, the maturing of unmanned but not unpiloted "drones," and the emergence of the cyber domain for conflict, have sparked both overenthusiastic and unduly critical commentaries. We should know that the feasible scope for the strategic effect of drones and of cyber is certain to be limited by what the classics ought to be permitted to tell us about the adversarial nature of strategy and about the perennial political strength of human social motivation to balance power presumed to be potentially hostile.

NOTES

INTRODUCTION. AIRPOWER AND STRATEGY

1. John Keegan, "Please Mr Blair, Never Take Such A Risk Again," *Daily Telegraph* (London), June 6, 1999.
2. Richard J. Overy, "Introduction," in *Air Power History: Turning Points from Kitty Hawk to Kosovo,* ed. Sebastian Cox and Peter Gray (London: Frank Cass, 2002).
3. For scholarly research on airpower, the following twelve books offer a point of departure: James Corum and Wray Johnson, *Airpower in Small Wars: Fighting Insurgents and Terrorists* (Lawrence: University Press of Kansas, 2003); Sebastian Cox and Peter Gray, eds., *Air Power History: Turning Points from Kitty Hawk to Kosovo* (London: Frank Cass, 2002); Colin S. Gray, *Airpower for Strategic Effect* (Maxwell AFB, Ala.: Air University Press, 2012); Richard P. Hallion, *Storm over Iraq: Air Power and the Gulf War* (Washington, D.C.: Smithsonian Institution Press, 1992); Benjamin S. Lambeth, *The Transformation of American Air Power* (Ithaca, N.Y.: Cornell University Press, 2000); Richard A. Mason, *Air Power: A Centennial Appraisal* (London: Brassey's, 1994); Phillip S. Meilinger, ed., *The Paths of Heaven: The Evolution of Airpower Theory* (Maxwell AFB, Ala.: Air University Press, 1997); John Andreas Olsen, ed., *A History of Air Warfare* (Washington, D.C.: Potomac Books, 2010); John Andreas Olsen, ed., *Global Air Power* (Washington, D.C.: Potomac Books, 2011); John Andreas Olsen, ed., *Air Commanders* (Washington, D.C.: Potomac Books, 2012); Alan Stephens, ed., *The War in the Air: 1914–1994* (Fairbairn, Australia: Air Power Studies Centre, 1994); and Martin van Creveld, *The Age of Airpower* (New York: Public Affairs, 2011).
4. See for example Robert A. Pape, *Bombing to Win: Air Power and Coercion* (Ithaca, N.Y.: Cornell University Press, 1996).
5. David S. Fadok, *John Boyd and John Warden: Air Power's Quest for Strategic Paralysis* (Maxwell AFB, Ala.: Air University Press, 1995). For more details on the men and their times, see James Coram, *Boyd: The Fighter Pilot Who Changed the Art of War* (Boston: Little, Brown, 2002); Grant T. Hammond, *The Mind of War: John Boyd and American Security* (Washington, D.C.: Smithsonian Institution Press, 2001); Frans Osinga, *Science, Strategy and War:*

The Strategic Theory of John Boyd (London: Routledge, 2007); John Andreas Olsen, *John Warden and the Renaissance of American Air Power* (Washington, D.C.: Potomac Books, 2007); and Richard T. Reynolds, *Heart of the Storm: The Genesis of the Air Campaign Against Iraq* (Maxwell AFB, Ala.: Air University, 1995).

6. Alan Stephens and Nicola Baker, *Making Sense of War: Strategy for the 21st Century* (Melbourne: Cambridge University Press, 2006), 13.

7. Michael Howard, *Causes of War and Other Essays* (London: Counterpoint, 1983), 215–16.

8. U.S. Army and U.S. Marine Corps, Field Manual 3-24 and Marine Corps Warfighting Publication 3-33.5, *Counterinsurgency* (Chicago: University of Chicago Press, 2007).

Chapter 1. Paradigm Lost

1. The wars include the prolonged Hapsburg-Valois conflicts, the Thirty Years War, the wars of Louis XIV, the wars of the French Revolution and Napoleon, the "Long War" of 1914–90, and the recent "War on Terrorism." See Philip Bobbitt, *The Shield of Achilles: War, Peace, and the Course of History* (New York: Alfred A. Knopf, 2002).

2. See Charles W. Shrader, "The Influence of Vegetius' *De re militari*," *Military Affairs* 45 (December 1981): 167.

3. See Azar Gat, *The Origins of Military Thought: From the Enlightenment to Clausewitz* (Oxford: Clarendon Press, 1989), 8.

4. For a revisionist and arguable response to this traditional interpretation of Machiavelli's thought, see Timothy R. W. Kubik, "Is Machiavelli's Canon Spiked? Practical Readings in Military History," *The Journal of Military History* 61 (January 1997): 7–30.

5. For example, see Frederick the Great, *Die Instruktion Friedrichs des Grossen für seine Generale von 1782*; Henry Humphrey Evans Lloyd, *The History of the Late War in Germany* (1766) and *Military Memoirs* (1781); and Adam Heinrich Dietrich von Bülow, *Der Geist des neuren Kriegssystems* (1799).

6. Quoted in Maj. Edward S. Johnston, "A Science of War," *The Command and General Staff Quarterly* XIV (June 1934): 90.

7. To characterize Newton as an irremediably linear and mechanistic scientist is unfair. As Barry Watts rightly points out, it was overzealous disciples such as Roger Boscovich (*A Theory of Natural Philosophy Reduced to a Law of Actions Existing in Nature*, 1758) and Pierre Simon de Laplace (*Philosophical Essay on Probabilities*, 1814) who linked the idea of linear predictability with

eighteenth-century mathematical physics. Newton did emphasize causality and long-term patterns, but it was Boscovich, Laplace, and others who popularized the idea that nature was entirely stable, rigidly deterministic, and required no divine intervention to work properly. See Barry D. Watts, *Clausewitzian Friction and Future War*, Institute for National Strategic Studies McNair Papers, no. 52 (Washington, D.C.: National Defense University, October 1996), 108–12.

8. See R. David Smith, "The Inapplicability Principle: What Chaos Means for Social Science," *Behavioral Science* 40 (January 1995): 30. See also Michael Howard, *Clausewitz* (Oxford: Oxford University Press, 1983), 13.
9. According to John Shy, Jomini's tainted stature among current students of war is undeserved. He was more than a doltish, thick-witted foil to Carl von Clausewitz. He was an astute analyst of Napoleonic warfare who warned that his maxims, principles, and prescriptions were not holy writ. Nevertheless, they did form the irreducible core of what some have perhaps unfairly characterized as a "Betty Crocker" approach to war. See John Shy, "Jomini," in *Makers of Modern Strategy*, ed. Peter Paret, with Gordon Craig and Felix Gilbert, 143–85 (Princeton, N.J.: Princeton University Press, 1986).
10. Quoted from John Lewis Gaddis, "International Relations Theory and the End of the Cold War," *International Security* 17 (Winter 1992–93): 6.
11. Johnston quotes Lord Grey in "A Science of War," 104. For a one-sided portrait of Clausewitz as a proto-chaos theorist who totally rejected the determinism of Jomini and his fellow rationalists, see Alan Beyerchen, "Clausewitz, Nonlinearity, and the Unpredictability of War," *International Security* 17, no. 3 (Winter 1992): 59–90.
12. Carl von Clausewitz, *On War*, trans. and ed. Michael Howard and Peter Paret (Princeton, N.J.: Princeton University Press, 1976), 86.
13. Quoted in Antulio J. Echevarria, "Moltke and the German Military Tradition: His Theories and Legacies," *Parameters* XXVI (Spring 1996): 96; see also Peter Paret, *Clausewitz and the State: The Man, His Theories, and His Times* (Princeton, N.J.: Princeton University Press, 1976); and Charles Edward White, *The Enlightened Soldier: Scharnhorst and the Militärische Gesellschaft in Berlin, 1801–1805* (Westport, Conn.: Praeger Publishers, 1989).
14. Quoted in Herbert Rosinski, "Scharnhorst to Schlieffen: The Rise and Decline of German Military Thought," *Naval War College Review* XXIX (Summer 1976): 85.
15. See Helmuth von Moltke, "Doctrines of War," in *War*, ed. Lawrence Freedman, 220–21 (Oxford: Oxford University Press, 1994).
16. Beyerchen, "Clausewitz, Nonlinearity, and War," 86; see also Clausewitz, *On War*, 149.
17. Clausewitz, *On War*, 85–87, 89, 148.

18. John Shy and Thomas W. Collier, "Revolutionary Warfare," in *Makers of Modern Strategy*, ed. Peter Paret, 843 (Princeton, N.J.: Princeton University Press, 1986).
19. Beyerchen, "Clausewitz, Nonlinearity, and War," 59. Clausewitz, of course, had his own limitations. For example, his vision of war did not adequately consider whether human violence, in addition to being a continuation of politics by other means, was also a cultural or biological activity. Further, Clausewitz's military romanticism gave limited thought to combined operations between land and sea; coalition warfare (the dominant form of war in the twentieth century); the technological and economic dimensions of war; and guerrilla warfare, revolutionary warfare, or military operations other than war.
20. Clausewitz, *On War*, 136.
21. Paul van Riper and Robert H. Scales Jr., "Preparing for War in the 21st Century," *Parameters* (Autumn 1997): 4–5.
22. The mediators include, for example, Julian Corbett, Ardant du Picq and, as already discussed, Clausewitz.
23. Members of the "Bomber Mafia" included Robert Olds, Kenneth Walker, Harold Lee George, Donald Wilson, Robert Webster, Laurence Kuter, Haywood Hansell, and Muir Fairchild. In turn, the COA and EOU were basically a mix of military planners and civilian economists.
24. Airpower theorists up through John Warden have traditionally minimized the interactive nature of air warfare, primarily because of their fixation on the "inherently offensive" nature of the medium. As a result, the defense has typically received short shrift in airpower theory. See Clausewitz, *On War*, 149.
25. Col. Thomas A. Fabyanic, USAF (Ret.), "War Doctrine, and the Air War College—Some Implications for the U.S. Air Force," *Air University Review* XXXVII (January–February 1986): 12.
26. Lt. Col. Don Wilson, "Long Range Airplane Development," Air Force Historical Research Agency (AFHRA), Maxwell AFB, Ala., file no. 248.211-17, November 1938, 5–6.
27. "Address of Major General Frank M. Andrews before the National Aeronautical Administration," January 16, 1939, AFHRA file no. 248.211–20, 8.
28. Nino Salvaneschi, "Let Us Kill the War: Let Us Aim at the Heart of the Enemy," 1917, AFHRA file no. 168.661-129, 31; Count Gianni Caproni, "Memorandum on 'Air War,'" 1917, AFHRA file no. 168.66-2, 2. Because Salvaneschi was an ardent popularizer of the emerging theories of Douhet and Caproni, especially in World War I, "Let Us Kill the War" does reflect their thinking at the time.
29. The examples appear in James C. Gaston, *Planning the American Air War: Four Men and Nine Days in 1941* (Washington, D.C.: National Defense University Press, 1982); Rick Atkinson, *Crusade: The Untold Story of the*

Persian Gulf War (New York: Houghton Mifflin Co., 1993); and Gen. William M. Momyer, *Air Power in Three Wars* (Washington, D.C.: U.S. Government Printing Office, 1978). The quotation is from Benjamin S. Lambeth, "Pitfalls in Force Planning: Structuring America's Tactical Air Arm," *International Security* 10 (Fall 1985): 92.

30. Caproni, "Memorandum on 'Air War,'" 2.
31. ACTS bomber instructor Muir "Santy" Fairchild was typical. He understood the dubious "logic" of metaphors but still subscribed to the "industrial web" theory of strategic bombardment. See Kenneth Schaffel, "Muir S. Fairchild: Philosopher of Air Power," *Aerospace Historian* 33 (Fall 1986): 167.
32. John Warden does try to infuse some flexibility into his theory. He admits that the individual importance and resiliency of his systems varies from one society (and one historical period) to another. He also agrees that it is extremely difficult to operate directly and successfully against single-leader states and organizations. Still, comparing closed societies to the human body is problematic.
33. For an excellent discussion of this relationship, see Mark Clodfelter, *Beneficial Bombing: The Progressive Foundations of American Airpower, 1917–1945* (Lincoln: University of Nebraska Press, 2010).
34. We can only assume that Spaatz, who believed in the paradigm-shifting power of the bomber, did not see the irony of rejecting the nascent nuclear age as a huge shift in its own right. After all, it ushered in an era of limited wars by limited means for limited ends that has yet to end. To confirm why this triumphalism was so obviously misplaced, see Richard Overy's judicious and sobering *The Bombing War: Europe 1939–1945* (New York: Penguin Books, 2013). The air war over Europe was indeed a grinding war of attrition, just like so many wars before it.
35. Comments by General John Pershing, n.d., 11, AFHRA file 248.211–16F.
36. See Robert A. Pape Jr., *Bombing to Win: Air Power and Coercion in War* (Ithaca, N.Y.: Cornell University Press, 1996); Maj. Thomas P. Ehrhard, USAF, *Making the Connection: An Air Strategy Analysis Framework* (Maxwell AFB, Ala.: Air University Press, 1995), 38.
37. Ehrhard, *Making the Connection: An Air Strategy Analysis Framework*, 50.
38. Ibid., 19.
39. Ibid., 50.
40. Consider, for example, the lack of fighter escorts and effective bombsights in the early phases of the Combined Bomber Offensive against Germany in World War II. Their absence, along with other limitations, sabotaged the revolutionary promise of a quick war dominated by aerial *Kesselschlachten* against the vital centers of the German state.

41. It is important to note that while punishment and denial strategies try to translate military effects into political change, decapitation strategies do the opposite. For a discussion of each approach, see Robert A. Pape Jr., "Coercion and Military Strategy: Why Denial Works and Punishment Doesn't," *The Journal of Strategic Studies* 15 (December 1992): 423–75.
42. This individual conclusion is obviously a limited and contentious one, but it is not really the point here. The central and appropriate one centers on picking the right type of strategy to secure your desired ends.
43. Pat A. Pentland, "Theater Strategy Development," unpublished manuscript in author's possession, 1993–94, 2–5.
44. Ibid., 3.
45. See Maj. Kevin E. Williams, *In Search of the Missing Link: Relating Destruction to Outcome in Airpower Applications* (Maxwell AFB, Ala.: Air University Press, 1994), 5–7.
46. See Carl Kaysen, *Note on Some Historic Principles of Target Selection* (Santa Monica, Calif.: RAND Corporation, Project RAND Research Memorandum 189 (RM-189), July 15, 1949).
47. The classification, however, is relative to the strategic situation and to the tactics and doctrine of your opponent. Ibid., 2.
48. Ibid., 4.
49. Ibid., 5.
50. Ibid., 5–6.
51. Salvaneschi, for example, claimed that the Allies "must aim, not at the army that fights, but at the factories of Essen." See Salvaneschi, "Let Us Kill the War," 38.
52. The classic case is the initial American belief that German machine tools, etc., used as many ball bearings as U.S. machines did. In fact, they did not, which meant that there was considerable slack in German ball-bearing stocks.
53. Irving L. Janis, *Air War and Emotional Stress* (Westport, Conn.: Greenwood Press, 1951), 87, 140, 143. See also Constantine Fitzgibbon, *The Winter of the Bombs* (New York: Norton, 1957); Hilton P. Goss, *Civilian Morale Under Aerial Bombardment 1914–1939*, 2 vols. (Maxwell AFB, Ala.: Documentary Research Division, Air University Libraries, Air University, 1948); Jack Hirshleifer, *Disaster and Recovery: A Historical Survey* (Santa Monica, Calif.: RAND Corporation Research Memorandum 3079 (RM-3079), 1963); Charles Ikle, *The Social Impact of Bomb Destruction* (Norman: University of Oklahoma Press, 1958); Hans Rumpf, *The Bombing of Germany*, trans. Edward Fitzgerald (New York: Holt, Rinehart and Winston, 1963); Richard M. Titmuss, *Problems of Social Policy: History of the Second World War, United Kingdom Civil Series* (London: Her Majesty's Stationery Office, 1950).

54. Although regular and unvarying air assaults do not increase civilian resentment against an attacker, irregular and variable assaults do the opposite. A logical question then follows: Is it worthwhile for an air planner to lower an opponent's ability to adapt (via sporadic air attacks) even though it raises his or her anger? Perhaps the question itself is too simple; if no active or passive defenses exist, or if demands for retaliation go unheeded, civilian hostility can shift back from the attacker to those domestic leaders who fail to provide organized support (adequate shelters, antiaircraft barrages, or relief measures) for their people. Since the aggression of a bombed population can be "diffuse and labile," and often equally directed at all sources of authority, *keeping an adversary off balance with irregular bombing may be a viable politico-military option, despite the possible rise in externally directed aggression it may inspire.* See Janis, *Air War and Emotional Stress*, 118, 127, 130, 135–137, and Alexander George, "Emotional Stress and Air War—A Lecture Given at the Air War College," November 28, 1951, 19, in AFHRA File No. K239.716251-65. For additional support that irregular bombing worked best against Germany, see K. W. Yarnold, *Lessons on Morale to be Drawn From Effects of Strategic Bombing on Germany: With Special Reference to Psychological Warfare*, Technical Memorandum, ORO-T-2 (Washington, D.C.: Operations Research Office, 1949).
55. Janis, *Air War and Emotional Stress*, 98, 103, 106–7, 144.
56. Ibid., 100.
57. George, "Emotional Stress and Air War," 12.
58. Ibid., 21.
59. The United States Strategic Bombing Survey (USSBS), despite the inconsistencies between its summary volumes and survey reports, largely supports Janis. One particular report concluded that continuous heavy bombing did not produce decreases in morale proportional to the amount of bombing carried out. It also determined that those who directly experienced the effects of air attack had much lower morale than those who experienced them indirectly. See United States Strategic Bombing Survey, *The Effects of Strategic Bombing on German Morale*, vol. I (Washington, D.C.: Government Printing Office, 1947), 33.
60. Janis, *Air War and Emotional Stress*, 127.
61. George, "Emotional Stress and Air War," 18.
62. Ibid.
63. See Stephen T. Hosmer, *Psychological Effects of U.S. Air Operations in Four Wars, 1941–1991* (Santa Monica, Calif.: RAND Corporation, 1996), xv–xxiii; Group Captain A. P. N. Lambert, RAF, "The Psychological Impact of Air Power Based on Case Studies since the 1940s" (MA thesis, Cambridge University, 1995), 79–91.

64. Obviously, this characterization of events is monocausal and requires a level of predictability that still eludes air strategists, despite the growing presence of precision weapons. See Maj. Jerry T. Sink, "Coercive Air Power: The Theory of Leadership Relative Risk," unpublished essay (Maxwell AFB, Ala.: USAF School of Advanced Airpower Studies, April 1993), 7–8; Ehrhard, *Making the Connection: An Air Strategy Analysis Framework*, 22–25.
65. Sink, "Coercive Air Power," 6–8. John Warden subsequently argued that the attack on the bunker had a positive strategic impact, but his assertion was highly controversial. Saddam's behavior, for example, may have been motivated by an increase in external risks and not by domestic considerations. He did, after all, allow Republican Guard units to suffer without appreciably modifying his behavior, and they were a major source of his political support.
66. Currently, however, in a fluid world where diplomatic conditions are more akin to those of the 1920s and 1930s, and where air technologies are no longer blunt and necessarily murderous instruments of war, a new generation of air planners has warmed to the idea of using aerospace power as an instrument of Schelling-like diplomacy.
67. Schelling defines the essence of bargaining as "the communication of intent, the perception of intent, the manipulation of expectations about what one will accept or refuse, the issuance of threats, offers, and assurances, the display of resolve and evidence of capabilities, the communication of constraints on what one can do, the search for compromise and jointly desirable exchanges, the creation of sanctions to enforce understandings and agreements, genuine efforts to persuade and perform, and the creation of hostility, friendliness, mutual respect, or rules of etiquette." Thomas C. Schelling, *Arms and Influence* (New Haven, Conn.: Yale University Press, 1966), 136.
68. Ibid., 143.
69. Ibid., 166.
70. Ibid., 134–35, 164.
71. Ibid., 142.
72. Ibid., 175.
73. Ibid., 180.
74. Ernest May, *Lessons of the Past* (New York: Oxford University Press, 1976), 134. That others dispute this monocausal explanation goes without saying, but their reservations do not invalidate May's argument as a "thought experiment" on the coercive use of force. For a more sophisticated analysis of how airpower contributed to the collapse of Italy, see Phil A. Smith, *Bombing to Surrender: The Contribution of Airpower to the Collapse of Italy, 1943* (Maxwell AFB, Ala.: Air University Press, 1998).

75. May, *The Lessons of the Past,* 137.
76. Ibid., 132, 134.
77. Ibid., 140–42. Thomas Fabyanic repeats May's conclusions, although more emphatically, in "Air Power and Conflict Termination," in *Conflict Termination and Military Strategy: Coercion, Persuasion, and War,* ed. Stephen J. Cimbala and Keith A. Dunn (Boulder, Colo.: Westview Press, 1993). According to Fabyanic, airpower destroyed, "beyond any doubt," the will of the Italians and the will and ability of the Japanese to wage war. In both instances, the loss of will then "brought about a change of government and new leaders who were not committed to a continuation of the war." Significantly, Fabyanic supports these bald assertions by citing *The Lessons of the Past,* that is, he depends upon the authority of another historian to "prove" his point. See p. 155.
78. Albert O. Hirschman, "The Search for Paradigms as a Hindrance to Understanding," *World Politics* 22 (April 1970): 339.
79. See Saburo Ienaga, *The Pacific War 1931–1945* (New York: Pantheon, 1978). After the war, Konoye portrayed himself as a patriot and anticommunist who quarreled repeatedly with the army. In fact, the Nuremburg trials may have inspired Konoye's new self-image.
80. George H. Quester, "The Impact of Strategic Air Warfare," *Armed Forces and Society* 4 (February 1978): 196.
81. Ibid.
82. May, *Lessons of the Past,* 142.
83. See Leon Sigal, *Fighting to a Finish: The Politics of War Termination in the United States and Japan, 1945* (Ithaca, N.Y.: Cornell University Press, 1988), 1–25; and Robert A. Pape, "Why Japan Surrendered," *International Security* 18 (Fall 1993): 154–201.
84. May cites Mussolini's Ethiopia campaign, the Sino-Japanese War, the Spanish Civil War, and the war in Vietnam to buttress his argument.
85. Kenneth Waltz, *Theory of International Politics* (Reading, Mass.: Addison-Wesley Publishing Co., 1979), 189 (emphasis added).
86. Ibid., 191.
87. Charles Glaser raises these questions in relation to nuclear weapons and our previous security dilemma with the former Soviet Union. The questions obviously apply to conventional air warfare as well. See Charles L. Glaser, *Analyzing Strategic Nuclear Policy* (Princeton, N.J.: Princeton University Press, 1990), 81–82.
88. See Sink, "Coercive Air Power: The Theory of Leadership Relative Risk," 2.
89. Ibid.

90. Ibid., 3.
91. See Maj. Martin L. Fracker, "Psychological Effects of Aerial Bombardment," *Airpower Journal* VI (Fall 1992): 56–67; George H. Quester, "The Psychological Effects of Bombing on Civilian Populations: Wars of the Past," in *Psychological Dimensions of War*, ed. Betty Glad, 201–14 (Newbury Park, Calif.: Sage Press, 1990).
92. Fracker, "Psychological Effects of Aerial Bombardment," 55.
93. Ibid., 59, 65. The attack (or campaign) must typically be massive and unrelenting.
94. Ibid., 59, 61–62.
95. As an analogy, think of the parallel campaigns conducted by General MacArthur and Admiral Nimitz in the Pacific Theater. At core, the campaigns represented competing paradigms on how to fight in the region, and the War Department, with the "arsenal of democracy" fully behind it, permitted both of them.
96. For a pioneering example of this sociological narrative, see Carl Builder, *The Icarus Syndrome: The Role of Air Power Theory in the Evolution and Fate of the U.S. Air Force* (New Brunswick, N.J.: Transaction, 1994). See also the Strategic Aerospace Warfare Study Panel, "Aerospace Power for the 21st Century: A Theory to Fly By," White Paper, Maxwell AFB, Ala., October 4, 1996. The latter distills the lost-in-the-institutional-wilderness argument as well as Builder does.
97. The idea of the American way of war was popularized by Russell Weigley and certainly applied at least up to the 1990s. Its basic beliefs could be characterized as follows: it is practical/utilitarian rather than abstract/theoretical; it is oriented toward problem solving; it is optimistic and can-do-oriented; it is technology-loving and technology-dependent; it is firepower-oriented; it is large-scale; it is aggressive and offensive; it is profoundly regular (in the sense of embracing Napoleonic-industrial warfare as its preferred paradigm of war); it is impatient; it is logistically excellent; it is highly sensitive to casualties; it is frequently tone deaf to the historical context it is operating in; it is frequently tone deaf to the cultural context it is operating in; it is only "flexibly"/weakly connected to political objectives; and finally, because of the previous three reasons, among others, American hard power has often been perceived as an *alternative* to diplomatic bargaining rather than an *extension* of it, or as a form of violent bargaining itself. For further details about this concept, see Russell F. Weigley, *The American Way of War: A History of United States Military Strategy and Policy* (Bloomington: Indiana University Press, 1977). For a thorough analysis of the doctrinal misalignments of the U.S. Army with the realities of Vietnam, see Andrew F. Krepinevich, *The Army and Vietnam* (Baltimore: Johns Hopkins University Press, 1986).

98. James G. Burton, *The Pentagon Wars* (Annapolis, Md.: Naval Institute Press, 1993), 44–45.
99. Boyd is anti-Jominian because his view of war is an "empty basket." There are no boilerplate axioms, prescriptions, or guidelines to clutter your thinking. What you are basically battling here is a process—decision making—rather than "stuff."
100. John A. Warden III, USAF, "The Enemy as a System," *Airpower Journal* IX (September 1995): 43.

Chapter 2. The Enemy as a Complex Adaptive System

1. For a longer discussion of his influence, as well as accolades for and critiques of his work, see Frans Osinga, *Science, Strategy and War, The Strategic Theory of John Boyd* (London: Routledge, 2007), ch. 1.
2. Colin Gray, *Modern Strategy* (Oxford: Oxford University Press, 1999), 90–91.
3. For more readable accounts of Boyd's life and work than my scholarly study of Boyd mentioned in note 1, see Robert Coram, *Boyd: The Fighter Pilot Who Changed the Art of War* (Boston: Little, Brown, 2002); and Grant Hammond, *The Mind of War: John Boyd and American Security* (Washington, D.C.: Smithsonian Press, 2001). For shorter introductions see, for instance, Frans Osinga, "John Boyd," in *Philosophers of War*, ed. Lee Eysturlid and John Coetzee, 203–12 (Santa Barbara, Calif.: Praeger, 2013); and Frans Osinga, "'Getting' a Discourse on Winning and Losing: A Primer on Boyd's Theory of Intellectual Evolution," *Contemporary Security Policy* 34, no. 3 (December 2013): 603–24. In addition to my book, this chapter draws on those two publications but expands on them significantly.
4. Gen. C. C. Krulak, Commandant of the Marine Corps, "Obituary," *Inside the Pentagon*, March 13, 1997, 5.
5. For other examples see, for instance, Phillip S. Meilinger, "Air Targeting Strategies: An Overview," in *Air Power Confronts An Unstable World*, ed. Richard Hallion, 60–61 (London: Brassey's, 1997), where the author states that "the key to victory was to act more quickly, both mentally and physically, than your opponent. He expressed this concept in a cyclical process he called the OODA Loop. As soon as one side acted, it observed the consequences, and the loop began anew. . . . The significance of Boyd's tactical air theories is that he later hypothesized that this continuously operating cycle was at play not only in a tactical aerial dogfight, but at the higher levels of war as well. In tracing the history of war Boyd saw victory consistently going to the side that could think the most creatively, and then acting quickly on that insight." This is not a critique of Meilinger, who merely includes Boyd as one of a several theorists

on airpower. But his rendering of Boyd's work is typical of most interpretations of Boyd. See also Phillip S. Meilinger, *Ten Propositions Regarding Air Power* (Washington, D.C.: Air Force History and Museums Program, 1995), 31–32, where he states that "John Boyd's entire theory of the OODA Loop is based on the premise that telescoping time—arriving at decisions or locations rapidly—is the decisive element in war because of the enormous psychological strain it places on an enemy." For similar brief and consequently limited discussions of the OODA loop, see Gary Vincent's two articles, "In the Loop, Superiority in Command and Control," *Airpower Journal* VI (Summer 1992): 15–25; and "A New Approach to Command and Control, the Cybernetic Design," *Airpower Journal* VII (Summer 1993): 24–38. See also Gordon R. Sullivan and James M. Dublik, "War in the Information Age," *Military Review* (April 1994): 47, where the authors lay out a vision of war in the information age. Incorporating the same pictogram of the OODA loop as used above, they argue that the concept of time has changed: "Tomorrow we will observe in real time, orient continuously, decide immediately and act within an hour or less." Remarkable is that Boyd is not listed as the intellectual father of the OODA loop, suggesting that the OODA construct has already become commonplace. The OODA loop has even been discussed in *Forbes* and *Harvard Business Review*. See Hammond, *The Mind of War*, 11.

6. See, for instance, Thomas Hughes, "The Cult of the Quick," *Air Power Journal* XV, no. 4 (Winter 2001): 57–68. Only in the endnotes does Hughes acknowledge that Boyd's ideas are more complex than this interpretation.
7. See the article by James Hasik, "Beyond the Briefing: Theoretical and Practical Problems in the Works and Legacy of John Boyd," *Contemporary Security Policy* 34, no. 3 (December 2013): 583–99.
8. See Jim Storr, "Neither Art Nor Science—Towards a Discipline of Warfare," *RUSI Journal* (April 2001): 39. Referring to Karl Popper, Storr states that "induction is unsafe" and "to generalize about formation-level C2 from aircraft design is tenuous."
9. Some of Boyd's briefings are available at http://www.ausairpower.net/APA-Boyd-Papers.html, and all are available at http://dnipogo.org/john-r-boyd/. All references to pages and slides in Boyd's work reflect the number written on Boyd's slides.
10. At http://dnipogo.org/john-r-boyd/, this presentation is included under the incorrect title Fast Transients.
11. John Boyd, "Abstract," in *A Discourse*, 1, http://dnipogo.org/john-r-boyd/.
12. Ibid.
13. Todd Stillman, "Introduction: Metatheorizing Contemporary Social Theorists," in *The Blackwell Companion to Major Contemporary Social Theorists*, ed.

George Ritzer, 3 (Oxford: Blackwell Publishing, 2003). See also Avi Kober, "Nomology vs Historicism: Formative Factors in Modern Military Thought," *Defense Analysis* 10, no. 3 (1994): 268. For the influence of formative factors on military thought, see also Azar Gat, *The Origins of Military Thought* (Oxford: Clarendon Press, 1989); *The Development of Military Thought: The Nineteenth Century* (Oxford: Clarendon Press, Oxford, 1992); and *A History of Military Thought, From the Enlightenment to the Cold War* (Oxford: Oxford University Press, 2001). The link between science and military thought is also explored in Robert P. Pellegrini, *The Links Between Science and Philosophy and Military Theory: Understanding the Past; Implications for the Future* (Maxwell AFB, Ala.: Air University Press, 1995); Barry Watts, *The Foundations of US Air Doctrine, the Problem of Friction in War* (Maxwell AFB, Ala.: Air University Press, 1984); and most recently by Antoine Bousquet, *The Scientific Way of War* (New York: Columbia University Press, 2009).

14. Stillman, "Introduction: Metatheorizing Contemporary Social Theorists," 3.
15. For a lengthy discussion of the formative factors of Boyd's work, see chapters 2, 3, and 4 in Osinga, *Science, Strategy and War*. Chapter 2 describes his personal experience and the organizational context of his work. Chapters 3 and 4 detail exhaustively how the academic and cultural zeitgeist influenced his work.
16. For an elaborate discussion of this doctrine innovation process see, for instance, Richard Lock-Pullan, *US Intervention Policy and Army Intervention* (London: Routledge, 2006).
17. For an in-depth description see chapters 3 and 4 of Osinga, *Science, Strategy and War*, on which this section is based, in particular chapter 3. See for a recent exploration of the influence of scientific developments on military thought, see Bousquet, *The Scientific Way of War*, in particular chapter 7, which draws from my book and which positions Boyd within the chaoplexic "regime," one of the four scientific "regimes" he distinguishes in the history of science. Lawrence Freedman, too, devotes considerable space to the influence of scientific ideas on Boyd's work. See his *Strategy: A History* (Oxford: Oxford University Press, 2013).
18. According to Peter Faber, "Boyd introduced the language of the New Physics, Chaos Theory, and Complexity Theory," in John Andreas Olsen, *Asymmetric Warfare* (Oslo: Royal Norwegian Air Force Academy Press, 2002), 58.
19. For an excerpt of Popper's 1975 lecture on evolutionary epistemology, see David Miller, *A Pocket Popper* (Oxford: Fontana, 1983), 78–86.
20. John Boyd, *The Conceptual Spiral*, 7, http://dnipogo.org/john-r-boyd.
21. Ibid.
22. John Boyd, *Organic Design for Command and Control*, 15, 18, http://dnipogo.org/john-r-boyd.

23. Ibid., 20.
24. One very readable account Boyd studied was, for instance, Howard Gardner, *The Mind's New Science: A History of the Cognitive Revolution* (New York: Basic Books, 1985). Other works he read that are considered part of the cognitive revolution include John von Neumann, *The Computer and the Brain* (New Haven, Conn.: Yale University Press, 1958); Norbert Wiener, *The Human Use of Human Beings: Cybernetics and Society* (Cambridge, Mass.: MIT Press, 1950); and Gilbert Ryle, *The Concept of Mind* (London: Hutchinson, 1949).
25. Gregory Bateson, *Mind and Nature: A Necessary Unity* (Cresskill, N.J.: Hampton Press, 2002).
26. Sergio Manghi, "Foreword," in Bateson, *Mind and Nature*, xi.
27. Bateson, *Mind and Nature*, 154.
28. In *Structuralism*, Piaget advanced the idea that there are "mental structures" that exist midway between genes and behavior. Mental structures build up as the organism develops and encounters the world. Structures are theoretic, deductive, a process. Interestingly, Piaget was influenced by Ludwig von Bertalanffy, among others. See Peter Watson, *A Terrible Beauty* (London: Phoenix Press, 2000), 629–30. Edward Hall, *Beyond Culture* (New York: Anchor Books, 1976), 42. This sentence appears on page 22 of Boyd's *The Strategic Game of ? and ?*.
29. Boyd, *Organic Design for Command and Control*, 13.
30. Ibid., 16.
31. John Boyd, *The Strategic Game of ? and ?*, 58, http://dnipogo.org/john-r-boyd/.
32. James Bryant Conant, *Two Modes of Thought* (New York: Trident Press, 1964), 31.
33. Ibid., 91.
34. In the bibliography of *Destruction and Creation* he also lists Maxwell Maltz, *Psycho-Cybernetics* (Hollywood, Calif.: Wilshire Book Company, 1971). The personal papers include, for instance, U. S. Anderson, *Success-Cybernetics: The Practical Application of Human-Cybernetics* (West Nyack, N.J.: Parker, 1970); F. H. George, *Cybernetics* (London: Hodder and Stoughton, 1971); Marvin Karlins and Lewis Andrews, *Biofeedback: Turning on the Power of Your Mind* (Philadelphia: J. B. Lippincott Co., 1973); and Norbert Wiener's *The Human Use of Human Beings: Cybernetics and Society* (New York: Avon Books, 1967).
35. Ludwig von Bertalanffy, *General Systems Theory* (New York: George Braziller, 1968), 141.
36. Ibid., 150. In popular systems thinking, this feature is enclosed in the notion of feed-forward, actions that are the result of expectation and anticipation.
37. Fritjof Capra, *The Web of Life, A New Scientific Understanding of Living Systems* (New York: Anchor Books, 1997).

38. Boyd, *The Strategic Game of ? and ?*, 28.
39. Boyd, *Organic Design for Command and Control*, 20. See also Boyd, *The Strategic Game of ? and ?*, 41.
40. John Boyd, *Patterns of Conflict*, 184, http://dnipogo.org/john-r-boyd. Note how Boyd uses the concept of friction not in the mechanical sense, as Clausewitz did, but in the thermodynamic sense, indicating that for Boyd friction refers to disorder.
41. The list included John Barrow's *Pi in the Sky, Counting, Thinking and Being* (New York: Back Bay Books, 1993), *The Artful Universe* (New York: Little, Brown & Company, 1996), and *Theories of Everything* (Oxford: Oxford University Press, 1991); Capra, *The Web of Life*; Peter Coveney and Roger Highfield's *Frontiers of Complexity* (New York: Ballantine Books, 1996), and *The Arrow of Time* (London: Flamingo, 1991); Richard Dawkins' *The Blind Watchmaker* (New York: W. W. Norton & Company, 1986), and *The Selfish Gene* (Oxford: Oxford University Press, 1976); Murray Gell-Mann, *The Quark and the Jaguar* (New York: Freeman & Company, 1994); Stephen Kellert, *In the Wake of Chaos* (Chicago: University of Chicago Press, 1993); Marvin Minsky, *Society of the Mind* (New York: Simon & Schuster, 1988); Robert Ornstein, *The Evolution of Consciousness* (New York: Prentice-Hall, 1991); Roger Penrose's *Shadows of the Mind: A Search for the Missing Science of Consciousness* (Oxford: Oxford University Press, 1994), and *The Emperor's New Mind: Concerning Computers, Minds and The Laws of Physics* (Oxford: Oxford University Press, 1989); Ilya Prigogine and Isabelle Stengers, *Order Out of Chaos* (London: Bantam Books, 1984); Stephen Rose, *The Making of Memory* (London: Bantam Press, 1992); David Ruelle, *Chance and Chaos* (Princeton, N.J.: Princeton University Press, 1991); Mitchell Waldrop, *Complexity: The Emerging Science at the Edge of Order and Chaos* (London: Viking, 1993).
42. Gell-Mann, *The Quark and the Jaguar*, 17.
43. Gell-Mann actually includes a military illustration here. See Gell-Mann, *The Quark and the Jaguar*, 293.
44. See, for instance, Michael Lissack, "Complexity: The Science, its Vocabulary, and its Relation to Organizations," *Emergence* 1, iss. 1 (1999): 110–26. But authors such as Ilya Prigogine already made the connection in the early 1980s, as did Charles Perrow in his seminal works *Normal Accidents* (Princeton, N.J.: Princeton University Press, 1999) and *Complex Organizations: A Critical Essay* (Glenview, Ill: Scott, Foresman, 1972). The literature on complexity theory and its relevance for the humanities, social sciences, and management theory is burgeoning. See, for instance, Shona L. Brown and Kathleen M. Eisenhardt, *Competing on the Edge, Strategy as Structured Chaos* (Cambridge, Mass.: Harvard Business School Press, 1998); Uri Merry, *Coping with Uncertainty: Insights from the New Sciences of Chaos, Self Organization, and Complexity* (Westport, Conn.: Praeger, 1995);

Raymond A. Eve et al., *Chaos, Complexity, and Sociology* (London: Sage, 1997); Kathleen Eisenhardt and Donald N. Sull, "Strategy as Simple Rules," *Harvard Business Review* (January 2001): 107–16; Eric D. Beinhocker, "Robust Adaptive Strategies," *Sloan Management Review* (Spring 1999): 95–106; Michael Church, "Organizing Simply for Complexity: Beyond Metaphor Towards Theory," *Long Range Planning* 32, no. 4 (1999): 425–40.

45. Russ Marion and Josh Bacon, "Organizational Extinction and Complex Systems," *Emergence* 1, no. 4 (1999): 76.
46. Ibid.
47. Susanne Kelly and May Ann Allison, *The Complexity Advantage, How the Sciences Can Help Your Business Achieve Peak Performance* (New York: McGraw-Hill, 1999), 5.
48. Henry Coleman, "What Enables Self-Organizing Behavior in Businesses," *Emergence* 1, iss. 1 (1999): 37.
49. Ibid., 40.
50. Pat Pentland, *Center of Gravity Analysis and Chaos Theory* (Maxwell AFB, Ala.: Air University Press, 1993). Interestingly, he also incorporates Boyd's OODA loop, acknowledging that this model and the essay, although developed in the 1970s, anticipated many of tenets of chaos theory and is consistent with it. Another example is Steven Mann, "Chaos Theory and Strategic Thought," *Parameters* XXII, no. 2 (Autumn 1992): 54–68. Writing in 1984, Barry Watts, in his book *The Foundations of US Air Doctrine: The Problem of Friction in War* (Maxwell AFB, Ala., 1984), explicitly acknowledges Boyd's influence in developing the publication.
51. Alan Beyerchen, "Clausewitz, Nonlinearity, and the Unpredictability of War," *International Security* 17, no. 3 (Winter 1992): 55–90.
52. John Boyd, *The Essence of Winning and Losing*, 5, http://dnipogo.org/john-r-boyd.
53. Boyd, *Patterns of Conflict*, 12.
54. John Boyd, *A New Conception for Air to Air Combat*, 19, http://dnipogo.org/john-r-boyd.
55. Ibid., 21.
56. Ibid.
57. Ibid., 22.
58. Ibid., 24.
59. Boyd, *Patterns of Conflict*, 134.
60. Ibid., 141.
61. Ibid., 111.

62. Ibid., 113.
63. Ibid., 98.
64. Ibid.
65. Ibid., 101.
66. Ibid., 99.
67. Ibid., 117.
68. Ibid.
69. Ibid., 137.
70. Ibid., 134.
71. Ibid.
72. Ibid., 141.
73. Ibid., 174.
74. Ibid., 176.
75. Ibid., 177.
76. Boyd, *The Strategic Game of ? and ?*, 58.
77. Ibid., 29.
78. Ibid., 33.
79. Ibid., 36
80. Ibid., 37.
81. Ibid., 47.
82. See Martin van Creveld, *Command in War* (Cambridge, Mass.: Harvard University Press, 1987).
83. Boyd, *The Strategic Game of ? and ?*, 45.
84. Ibid., 58.
85. Boyd, *Organic Design for Command and Control*, 12.
86. Boyd, *Patterns of Conflict*, 72.
87. See, for instance, Andrew Zolli and Ann Marie Healy, *Resilience: Why Things Bounce Back* (New York: Simon & Schuster, 2012); and Max McKeown, *Adaptability: The Art of Winning in an Age of Uncertainty* (London: Kogan Page, 2012).
88. See, for instance, his remarks in Conant, *Two Modes of Thought*, 78; Chilton Pearce, *The Crack in the Cosmic Egg* (New York: Pocket Books, 1974), 95; Thomas Kuhn, *The Structure of Scientific Revolutions* (Chicago: Chicago University Press, 1970), 64, 66, 86, and 162; and Michael Polanyi, *Knowing and Being* (London: Routledge and Kegan Paul, 1969), 155. Boyd's personal papers and his library can be found at the U.S. Marines Corps University Library.

89. Boyd, *The Essence of Winning and Losing*, 5.
90. Ibid., 2.
91. On the merit of metaphors, see, for instance, Gareth Morgan, *Images of Organizations* (New York: Sage, 1986).
92. Gray, *Modern Strategy*, 124–126, distinguishes four levels to categorize strategic theories: (1) A level that transcends time, environment, political and social conditions, and technology (for instance, Clausewitz and Sun Tzu); (2) A level that explains how the geographical and functional complexities of war and strategy interact and complement each other (Corbett on naval warfare); (3) A level that explains how a particular kind or use of military power strategically affects the course of conflict as a whole (Mahan, Douhet, Schelling on the role of maritime power, airpower, and nuclear power respectively); (4) A level that explains the character of war in a particular period, keyed to explicit assumptions about the capabilities of different kinds of military power and their terms of effective engagement (the use of airpower as a coercive tool).
93. David S. Fadok, "John Boyd and John Warden: Air Power's Quest for Strategic Paralysis," in *The Paths of Heaven*, ed. Phillip Meilinger (Maxwell AFB, Ala.: Air University Press, 1997).
94. Boyd, *Patterns of Conflict*, 174.
95. This sections draws from Frans Osinga, "Air Warfare," in Julian Lindley-French and Yves Boyer, *The Oxford Handbook of War* (Oxford: Oxford University Press, 2012).
96. Thomas A. Keaney and Eliot A Cohen, *A Revolution in Warfare, Air Power in the Persian Gulf* (Annapolis, Md.: Naval Institute Press, 1995), 188. For the view that the full maturing of air power in the 1980s and 1990s is the real RMA/MTR today, see for instance Benjamin Lambeth, "The Technology Revolution in Air Warfare," *Survival* 39, no. 1 (Spring 1997): 65–83.
97. Richard P. Hallion, *Storm over Iraq, Air Power and the Gulf War* (Washington, D.C.: Smithsonian Institution Press, 1992), 252.
98. The following short overview of the "lessons" of Desert Storm is based on several sources. Besides Keaney and Cohen's *A Revolution in Warfare*, and Hallion's *Storm Over Iraq*, see Michael R. Gordon and Gen. Bernard E. Trainor, *The General's War* (Boston: Little, Brown and Company, 1995). For a well-researched counterpoint on the alleged decisive impact on Iraqi ground units, see Daryl G. Press, "The Myth of Air Power in the Persian Gulf War and the Future of War," *International Security* 26, no. 2 (Fall 2001): 5–44.
99. Keaney and Cohen, *A Revolution in Warfare*, 205.
100. A growing point of concern thus became the high percentage of casualties caused by "friendly fire."

101. For a good discussion of this distinction, see Stephen Hosmer, *Psychological Effects of U.S. Air Operations in Four Wars, 1941–1991* (Santa Monica, Calif.: RAND, 1996), ch. 10.
102. James Titus, *The Battle of Khafji: An Overview and Preliminary Analysis* (Maxwell AFB, Ala.: Air University Press, 1996), 19.
103. Hallion, *Storm over Iraq*, 121–23.
104. For a balanced account of the strategic air offensive against Iraq, see Richard Davis, "Strategic Bombardment in the Gulf War," in *Case Studies in Strategic Bombardment*, ed. R. Cargill Hall (Washington, D.C.: U.S. Government Printing Office, 1998).
105. Joint Chiefs of Staff, *Joint Vision 2010* (Washington, D.C.: U.S. Department of Defense, 1997), 17. For a short description of *Joint Vision 2010*, see Maj. Gen. Charles Link, "21st Century Armed Forces—Joint Vision 2010," *Joint Forces Quarterly* (Autumn 1996): 69–73.
106. David Alberts, *Information Age Transformation, Getting to a 21st Century Military* (Washington, D.C.: U.S. Department of Defense, CCRP Publications, 1996), 7.
107. David Gompert, Richard Kugler, and Martin Libicki, *Mind the Gap: Promoting a Transatlantic Revolution in Military Affairs* (Washington, D.C.: National Defense University Press, 1997), 4.
108. *DoD Report to Congress on NCW* (Washington, D.C.: U.S. Department of Defense, CCRP Publications, 2001), 3-1.
109. These "tenets" appear in several network-centric warfare publications. See *DoD Report to Congress on NCW*, i, v, or 3-10. See also David S. Alberts, John J. Gartska, and Frederick P. Stein, *Network Centric Warfare* (Washington, D.C.: U.S. Department of Defense, CCRP Publications, 1999), for an early coherent description of network-centric warfare with references to Boyd.
110. It is estimated that between 8,000 and 12,000 Taliban fighters were killed. Estimates of the numbers of Afghan civilian casualties vary from eight hundred to thirty-five hundred. See Michael O'Hanlon, "A Flawed Masterpiece," *Foreign Affairs* (May/June 2002): 49. See also Max Boot, "The New American Way of War," *Foreign Affairs* 82 (July/August 2003): 41–50; and Stephen Biddle, "Afghanistan and the Future of Warfare," *Foreign Affairs* 82 (March/April 2003): 31–48.
111. International Institute for Strategic Studies, *Strategic Survey 2001–2002* (London: IISS, 2002), 71.
112. In fact after several days 90 percent of all targets were emergent targets instead of planned targets.
113. For detailed analyses of Iraqi Freedom, see, for instance, Anthony Cordesman, *The Iraq War: Strategy, Tactics, and Military Lessons* (Westport, Conn.: CSIS,

Praeger, 2003). Michael Gordon and Bernard Trainor, *Cobra II: The Inside Story of the Invasion and Occupation of Iraq* (New York: Random House, 2006), offers a good insight into the ground war. For a welcome study of the air operations of Iraqi Freedom, see Benjamin Lambeth, *The Unseen War: Allied Air Power and the Takedown of Saddam Hussein* (Annapolis, Md.: Naval Institute Press, 2013).

114. For operations in North Iraq, see Richard Andres, "The Afghan Model in Northern Iraq," in *War in Iraq: Planning and Execution*, ed. Thomas Keaney and Thomas Mahnken, 52–64 (London: Routledge, 2007). For the effectiveness of air interdiction, see Richard Andres, "Deep Attack against Iraq," in *War in Iraq: Planning and Execution*, ed. Thomas Keaney and Thomas Mahnken, 69–88 (London: Routledge, 2007). For the psychological impact of the air threat, see Kevin Woods, "Doomed Execution," in *War in Iraq: Planning and Execution*, ed. Thomas Keaney and Thomas Mahnken, 97–123 (London: Routledge, 2007).

115. For a discussion of this in the context of drone warfare against the Taliban and Al-Qaeda, see a summary of recent insights in Frans Osinga, "Bounding the Debate on Drone Warfare: The Paradox of Post-Modern War," in Herman Amersfoort, Desiree Verweij, and Joseph Soeters, *Moral Responsibility and Military Effectiveness* (Amsterdam: Asser Press, 2013).

116. See Karl Mueller, ed., *Precision and Purpose: Airpower in the Libyan Civil War* (Santa Monica, Calif.: RAND, 2014).

117. For a positive view on the Western way of warfare, see Christopher Coker, *Humane Warfare* (London: Routledge, 2002); and Andrew Latham, "Warfare Transformed: A Braudelian Perspective on the 'Revolution in Military Affairs,'" *European Journal of International Relations* 8, no. 2 (2002). For a critical perspective, see Martin Shaw, *The New Western Way of War* (Cambridge: Polity Press, 2005). See also Michael Ignatieff, *Virtual War, Kosovo and Beyond* (London: Chatto and Windus, 2000); Colin McInnes, *Spectator-Sport War* (London: Rienner, 2002); Mikkel Rasmussen, *The Risk Society at War* (Cambridge: Cambridge University Press, 2007); and Christopher Coker, *War in the Age of Risk* (Cambridge: Polity, 2009).

118. For this conclusion, see the following excellent articles: Barry Posen, "The War for Kosovo: Serbia's Political-Military Strategy," *International Security* 24, no. 4 (Spring 2000): 39–84; Daniel L. Byman and Matthew C. Waxman, "Kosovo and the Great Air Power Debate," *International Security* 24, no. 4 (Spring 2000): 5–38; Andrew L. Stigler, "A Clear Victory for Air Power: NATO's Empty Threat to Invade Kosovo," *International Security* 27, no. 3 (Winter 2002/03): 124–57; Frank Harvey, "Getting NATO's Success in Kosovo Right: The Theory and Logic of Counter-Coercion," *Conflict Management and Peace Science* 23 (2006): 139–58.

119. See Osinga, *Science, Strategy and War,* 106–11 and 242–43.
120. Boyd, *Patterns of Conflict,* 124–25.

CHAPTER 3. SMART STRATEGY, SMART AIRPOWER

1. Ferdinand Foch reportedly said this in approximately 1911 while serving as a professor of strategy at France's Ecole Superieure de Guerre, http://www.greatquotes.com/quote/861686.
2. "Military Aircraft: Early History," Encyclopedia Britannica Academic Edition, http://www.britannica.com/EBchecked/topic/382295/military-aircraft/57483/Early-history?anchor=ref521642, accessed November 23, 2010.
3. Carl von Clausewitz, *On War,* rev. ed., trans. and ed. Michael Howard and Peter Paret (Princeton, N.J.: Princeton University Press, 1976), 97.
4. "Nation" as used in this chapter means any organization that intends to use some kind of force to achieve its objectives. Regardless of its form (state or terrorist organization) it needs to accomplish the same tasks.
5. Ideally, there would be yet a third future picture: one for the whole world or at least for the region where the conflict might take place. The process for developing global or regional future pictures is the same as for your own country or group and for the opponent.
6. For simplicity, we will frequently call the planning entity a "country." The strategy concepts, however, apply equally to a country, a group, an alliance, or a unit. This is not a nation-state-specific strategy methodology!
7. Maurice de Saxe, *Reveries on the Art of War,* trans. and ed. by Brig. Gen. Thomas R. Phillips (Mineola, N.Y.: Dover Publications, Inc., 2007), 121.
8. Sun Tzu, *The Art of War,* translated from the Chinese by Lionel Giles, M.A., 1910 (Hollywood, Fla.: Simon & Brown, 2010), 11.
9. For a superb history of the Byzantine Empire and the brilliance of its policy, see Edward N. Luttwak, *The Grand Strategy of the Byzantine Empire* (Cambridge, Mass.: Belknap Press of the Harvard University Press, 2009).
10. Take, for example, an entity like Al-Qaeda. If all it had were the strong desire to kill nonbelievers, it would be little more than an academic curiosity and its P_S of changing its opponents would be zero. Only when it acquires physical capabilities like money, communications, pamphlets, schools, pilots, and stolen aircraft can it raise its P_S above zero. Note that an entity only needs to have physical assets at its disposal; it does not need to own them in the way that most nations own their physical assets.
11. United States Strategic Bombing Survey, *Summary Report (European War)* (Washington, D.C.: United States Government Printing Office, 1946), 16, para

4. "Their morale, their belief in ultimate victory or satisfactory compromise, and their confidence in their leaders declined, but they continued to work efficiently as long as the physical means of production remained."

12. In retrospect, I believe I overstated the importance of the strategic psychological operations, which were not executed for a variety of reasons. I believe that if they had been executed as proposed, that Saddam would have been overthrown, which would have been good for Iraq and the world. But since his overthrow was not essential, it was not logical to say that the strategic psychological operations were as important as the bombing operations. It would have been far more correct to have said that the strategic psychological operations had the possibility of achieving significant results for very little cost and that it would be a huge error not to try them as long as we kept in mind that they were unpredictable and had to be subordinate to the physical operations.
13. In simple form there were four objectives: Iraq out of Kuwait, restoration of the Kuwaiti government, safety for Americans in the area, and a more stable region (meaning a less powerful Iraq).
14. For more detail see Albert-László Barabási and Eric Bonabeau, "Scale-Free Networks," *Scientific American,* May 2003, 60.
15. The second ring has experienced several name changes since my first draft of the concept before the first Gulf War. I originally called it "key production" but came to realize that people were translating the idea as "manufacturing," which was not at all the idea. I then called it "organic essentials" to capture the idea that there were processes necessary for a system to function properly. That name did not work because some people thought that "organic" meant agriculture. I have most recently adopted the simple name "processes" and have found this word to work satisfactorily for both military and business applications. When you think about processes, think about conversion mechanisms such as electrical generation, communications, recruiting, etc. In the original version of the five rings, I called the fifth ring "fielded forces" because I was only concerned at the time with geopolitical structures and the name worked well. As I subsequently took the five rings into the business world, I found that "fielded forces" was confusing so I changed the name to "agents" and subsequently to "action units." "Action units" is a broader term and is preferable to "fielded forces" outside the military. Users who are only interested in the military application of the five rings can certainly use the older, more limited term, as long as they do not lose sight of the fact that there may be fifth-ring center of gravity targets in an opponent's system that are not part of the military (a civilian hospital that you may want to protect and improve, for example).
16. At the time of the first Gulf War, we included communications in the first ring. After a lot of thought and experience in using the model in many other places, however, it became clear that communications was not just the province of the

leader, but an element that affected everyone in the system all the time. Thus the decision to put it in the second ring.

17. Attacks on the system may have a big impact on the enemy leadership. If the leadership is rational, it is likely to sue for peace before its system is paralyzed or destroyed (for example, the Serbs in the 1999 war). The leadership will generally assess the cost of rebuilding, the effect on the state's economic position in the postwar period, the internal political effect on their own survival, and whether the potential gain from continuing the war is worth the cost. It is normally (but not always) an excellent outcome for you when the enemy leadership makes the right decision prior to your completing your operations. But again, you should do your best to avoid dependence on a rational decision. See earlier discussion about psychological operations.
18. Sir B. H. Liddell Hart, *Strategy* (New York: Meridian, 1991), 204.
19. Noel Barber, *The War of the Running Dogs: How Malaya Defeated the Communist Guerrillas 1948–60* (London: Cassel, 2004). See especially "The Briggs Plan," 116–18.
20. An impact plan expresses the concept of an "effect" as in effect-based operations but is specific in that it has an end state (desired effect), measure of success, and time frame for when the end state (effect) must be realized.
21. Sun Tzu, *The Art of War*, 8.
22. RAF Bomber Command was quite active during 1943 but virtually all of its attacks were at night and directed against cities with a major objective of "lowering morale and de-housing workers." Richard G. Davis, *Bombing the European Axis Powers: A Historical Digest of the Combined Bomber Offensive, 1939–1945* (Maxwell AFB, Ala.: Air University Press, 2006). Data extracted from a CD in the back of that book that contains the attached files "1943.xls" and "I Sheet Key.pdf." Quote is from the 1 Sheet Key pdf page 6 in a section titled General Information.
23. Ibid. For purposes of this analysis, I counted as an "attack" only missions that started with ten or more aircraft. Of the approximately seventy-two missions of fewer than ten aircraft, twenty-nine hit undefined industrial areas, and thirty ended up being against undefined targets of opportunity, which in 1943 generally meant that the mission was unable to find its assigned targets due to weather so it dropped on what it could find from altitude. The eleven target categories hit in 1943 are as follows with my assessment in parenthesis of the "ring" in which the target would have fallen using the five-ring system: aircraft manufacturing (5); bearings (2); industrial areas (3 or 4); marshaling yards (2); synthetic oil plants (2); port areas (5); shipping (5); steel (2); rubber (2); tires (5); and U-boat shipyards (5). As an aside, if I were planning a similar attack against a similar opponent today, I would not hit steel, industrial areas,

or rubber as they are not really ring-2 targets because they do not have a general impact on the enemy as a system.

24. Global Security.org, Operation Allied Force, http://www.globalsecurity.org/military/ops/allied_force.htm.
25. I use the word domain in this context to illustrate the dramatic difference between the world of serial and parallel attacks. The USAF does not refer to serial and parallel domains but rather uses the word in connection with air, space, and cyber.

CHAPTER 4. FIFTH-GENERATION STRATEGY

1. For commentary on this relationship, see Niall Ferguson, *Civilization: The West and the Rest* (London: Allen Lane, 2011).
2. Carl von Clausewitz, *On War*, ed. Anatol Rapoport (Harmondsworth: Penguin, 1982).
3. See Phillip S. Meilinger, "Busting the Icon: Restoring Balance to the Influence of Clausewitz," *Strategic Studies Quarterly* (Fall 2007); and Nikolas Gardner, "Resurrecting the 'Icon': The Enduring Relevance of Clausewitz's *On War*," *Strategic Studies Quarterly* (Spring 2009).
4. Michael Howard, "Narratives of War," *The Times Literary Supplement*, April 3, 2013. For an unusually clear discussion of the trinity, see Christopher Bassford, "Teaching the Clausewitzian Trinity," http://www.clausewitz.com/readings/Bassford/Trinity/TrinityTeachingNote.htm, accessed October 6, 2013. See also Clausewitz, *On War*, 101–22.
5. For example, Gardner, "Resurrecting the 'Icon,'" 121; and Peter Paret, "Clausewitz," in *Makers of Modern Strategy*, ed. Peter Paret, 206–7 (Princeton, N.J.: Princeton University Press, 1986).
6. Gardner, "Resurrecting the 'Icon,'" 121.
7. For commentary on strategy in the South Pacific, see Alan Stephens, "George C. Kenney: 'A Kind of Renaissance Airman,'" in *Air Commanders*, ed. John Andreas Olsen, 66–100 (Washington: Potomac Books, 2012).
8. American involvement in Vietnam began in the 1950s and escalated throughout most of the 1960s. U.S. combat forces withdrew in 1973. For two of the better commentaries, see Stanley Karnow, *Vietnam: A History* (New York: Penguin Books, 1984); and Neil Sheehan, *A Bright Shining Lie* (London: Jonathan Cape, 1989).
9. Karnow, *Vietnam*, 694–98.
10. For an insight into this kind of unforeseen effect, see Kenneth J. Hagan and Ian J. Bickerton, *Unintended Consequences: The United States at War* (London: Reaktion Books, 2007).

11. During the surge about 20,000 additional soldiers were deployed to Iraq and the tours of some units already in-country were extended.
12. Andrew J. Bacevich, "Surge to Nowhere," *Washington Post,* January 20, 2008.
13. "Secret Iraq—Awakening," *Four Corners,* Australian Broadcasting Commission Television, October 18, 2010.
14. Ibid.
15. U.S. Army and U.S. Marine Corps, Field Manual 3-24 and Marine Corps Warfighting Publication 3-33.5. *Counterinsurgency.* Chicago: University of Chicago Press, 2007.
16. Andrew J. Bacevich, *Washington Rules: America's Path to Permanent War* (New York: Metropolitan Books, 2010), 196–202.
17. Joshua Partlow, "Karzai Wants US to Reduce Military Operations in Afghanistan," *Washington Post,* November 14, 2010.
18. Rod Nordland, "Troop 'Surge' in Afghanistan Ends With Mixed Results," *New York Times,* September 21, 2012. After the withdrawal, 68,000 U.S. troops remained in Afghanistan.
19. Ibid.
20. For a perceptive early exposition of this argument, see Celeste Ward Gventer, "A False Promise of 'Counterinsurgency,'" *New York Times,* December 1, 2009.
21. The notion of the three-block war was promoted in the late 1990s by the then-commandant of the U.S. Marine Corps, Gen. Charles C. Krulak. Charles C. Krulak, "The Strategic Corporal: Leadership in the Three Block War," *Marines Magazine* (January 1999). See also Max Boot, "Beyond the 3-Block War," *Armed Forces Journal* (March 2006); and Gen. John Abizaid, "Combined Civil/Military Responses to National and International Events," The Future Australian Defence Force: Learning from the Past, Planning for the Future, Canberra, Australian Defence College and Royal United Services Institute Seminar 2007, May 16, 2007.
22. Robert O'Neill, "Restoring Utility to Armed Force in the 21st Century," Strategic and Defence Studies Centre 40th Anniversary Seminar Series, Canberra, Australian National University, August 15, 2006.
23. Quoted in Bob Woodward, *Obama's Wars* (New York: Simon & Schuster, 2010), 259.
24. Emile Simpson, *War from the Ground Up: Twenty-First Century Combat as Politics* (London: C. Hurst & Co., 2012).
25. Emile Simpson, "The Fusion of Military and Political Activity in 21st Century Combat," a presentation at the Strategic and Defence Studies Centre, Australian National University, Canberra, October 18, 2013. Simpson notes that in a kaleidoscopic environment there will be marginal players who do not represent a

political threat, who should not be confused with the "real" enemy, and against whom military resources should not be diverted. Examples include narcotics gangs, criminals, pirates, profiteers, and the like.

26. Other generations are represented by the F-102 and F-106 (2nd), the F-4 (3rd), and the F-15, F-16, and F-18 (4th).
27. Parts of the remaining section of this chapter first appeared in Alan Stephens and Nicola Baker, *Making Sense of War: Strategy for the 21st Century* (Melbourne: Cambridge University Press, 2006) and are used here by permission.
28. Examples include the Doolittle raid against Japan in 1942, the Israeli Air Force's strikes against Iraq's Osirak nuclear reactor in 1981 and Libya's nuclear reactor near Deir ez-Zor in 2007; and Al-Qaeda's attack on the World Trade Center in New York City in 2001.
29. *The Book of Ser Marco Polo, the Venetian, Concerning the Kingdom and Marvels of the East,* ed. Henry Yule (London: John Murray, 1903).
30. Christopher Clark, "The First Calamity," *The London Review of Books*, August 29, 2013, 5.
31. Stephens, "George C. Kenney: 'A Kind of Renaissance Airman.'"
32. See Richard P. Hallion, *Storm over Iraq* (Washington, D.C.: Smithsonian Institution Press, 1992); Robert C. Owen, ed., *Deliberate Force: A Case Study in Effective Air Campaigning* (Maxwell AFB, Ala.: Air University Press, 2000); Benjamin S. Lambeth, *The Transformation of American Air Power* (Ithaca, N.Y.: Cornell University Press, 2000); and Benjamin S. Lambeth, *The Unseen War: Allied Air Power and the Takedown of Saddam Hussein* (Annapolis, Md.: Naval Institute Press, 2013).
33. John A. Warden III, *The Air Campaign* (Washington, D.C.: Brassey's, 1989). Warden has elaborated on his model in a number of journal articles. See also John Andreas Olsen, *John Warden and the Renaissance of American Air Power* (Washington, D.C.: Potomac Books, 2007).
34. Hallion, *Storm over Iraq*, 262–65, 303–7.
35. This lesson at least seems to have been learned. Commenting on plans for a possible strike against the Syrian regime of President Bashar al-Assad in September 2013, chairman of the U.S. Joint Chiefs of Staff, Gen. Martin Dempsey of the Army, noted that "punitive strikes . . . do not translate so simply into military target sets." "The Tomahawks Fly," *The Economist*, August 31, 2013.
36. Boyd never published a substantive work on strategy. His thinking was formalized in several large slide shows, built up over years and which could take two or more days to present. The most important of these was his *Discourse on Winning and Losing.* For more on Boyd, see Grant T. Hammond, *The Mind of War: John Boyd and American Security* (Washington, D.C.: Smithsonian

Institution Press, 2001); and Robert Coram, *Boyd: The Fighter Pilot Who Changed the Art of War* (Boston: Little, Brown, 2002).

37. Sun Tzu, *The Art of War*, ed. Samuel B. Griffith (Oxford: Oxford University Press, 1963).

38. The 10-to-1 ratio was claimed during the Korean War, but subsequently the figure was found to be closer to 2-to-1, perhaps even less. Ironically, the difference, while extreme, does not invalidate Boyd's associated deduction, which relates to the decision-making contest generally rather than to the F-86/MiG contest per se.

39. David Fadok also used "form" to describe Warden's focus but chose "process" instead of "purpose" for Boyd. See David S. Fadok, *John Boyd and John Warden: Air Power's Quest for Strategic Paralysis*, School of Advanced Air Power Studies (Maxwell AFB, Ala.: Air University Press, 1995).

40. See Hallion, *Storm over Iraq;* and Lambeth, *The Unseen War.*

41. R. J. Overy, *The Air War 1939–1945* (London: Papermac, 1987), 54.

42. My thanks to John Andreas Olsen for this insightful observation. E-mail to author, September 25, 2013.

43. Bremer was administrator in Iraq from May 2003 to June 2008. His major errors were to ban all members of the Baath Party (Saddam Hussein's power base) from holding office in the new Iraqi civil service, thus depriving it of most of its expertise, and to disband the Iraqi army, thus forcing large numbers of trained, angry young men onto the streets. Thomas E. Ricks, *Fiasco: The American Military Adventure in Iraq* (London: Penguin Books, 2007), 158–66.

44. Clausewitz, *On War*, 119.

45. Robert L. Scales, "Checkmate by Operational Maneuver," *Armed Forces Journal International* (October 2001).

46. Ibid.

47. Gen. Richard Myers, quoted in "US Push to Base Forces on our Soil," *The Weekend Australian*, January 17, 2004, 001.

48. For example, no reference to the concept of checkmate by operational maneuver was found in a survey of the professional journals of the U.S. Army and the Australian army from 2002 to 2013. See *Parameters, The US Army War College Quarterly* (Carlisle, Pa.: Strategic Studies Institute); and *Australian Army Journal* (Duntroon: Land Warfare Studies Centre).

49. Elaine M. Grossman, "The Halt Phase Hits a Bump," *Air Force Magazine* (April 2001).

50. Thom Shanker and Eric Schmitt, "Pentagon Weighs Strategy Change to Deter Terror," *New York Times*, July 5, 2005.

51. Walter J. Boyne, *Operation Iraqi Freedom* (New York: Tom Doherty Associates, 2003), 92.
52. Patrick Cockburn, "What We Haven't Heard About Iraq," *London Review of Books* (October 10, 2013), 38–39.
53. See Brian McAllister Linn, *The Echo of Battle: The Army's Way of War* (Cambridge, Mass.: Harvard University Press, 2007).
54. For commentary on Powell and Schwarzkopf, see David Halberstam, *War in a Time of Peace* (London: Bloomsbury, 2001), 47, 51; and Eliot A. Cohen, *Supreme Command* (New York: The Free Press, 2002), 190–91; for Clark, see Benjamin S. Lambeth, *NATO's Air War for Kosovo* (Santa Monica, Calif.: RAND, 2001), 199–204; for Franks, an original assessment is provided by a senior Chinese PLA officer, Lieutenant General Liu Yazhou, "Interview with Lieutenant General Liu Yazhou," in *Heartland: Eurasian Review of Geopolitics* (Hong Kong: Gruppo Editoriale, L'Esspresso/Cassan Press, January 2005). See also Stephen Budiansky, "Of Tools and Tasks: Air War—Striking in Ways We Haven't Seen," *Washington Post*, April 26, 2003, B01.
55. See Sean Naylor, *Not a Good Day to Die: The Untold Story of Operation Anaconda* (New York: Berkley Caliber, 2005). Anaconda was planned by the Army with little consultation with the Navy, Marines, and Air Force, who would have to provide most of the fire support.
56. Ricks, *Fiasco: The American Military Adventure in Iraq*, 127–29.
57. Owen, *Deliberate Force: A Case Study in Effective Air Campaigning.*
58. Lawrence E. Harrison and Samuel P. Huntington, *Culture Matters* (New York: Basic Books, 2000); and Samuel P. Huntington, "The Clash of Civilisations?" *Foreign Affairs* (Summer 1993).
59. J. F. C. Fuller, *The Foundations of the Science of War* (Forth Leavenworth, Kan.: Combat Studies Institute Press, 2012).
60. Lewin (1890–1947) is widely regarded as the father of those fields of research in applied psychology.

Chapter 5. Airpower Theory

1. This chapter is based on Colin S. Gray, *Airpower for Strategic Effect* (Maxwell AFB, Ala.: Air University Press, 2012), ch. 9.
2. Antulio J. Echevarria II, *Clausewitz and Contemporary War* (Oxford: Oxford University Press, 2007), 7.
3. Harold R. Winton, "An Imperfect Jewel: Military Theory and the Military Profession," *The Journal of Strategic Studies* 34, no. 6 (December 2011): 854–56.

4. I am indebted to I. B. Holley, "Reflections on the Search for Airpower Theory," in *The Paths of Heaven: The Evolution of Airpower Theory,* ed. Philip S. Meilinger, 579–99 (Maxwell AFB, Ala.: Air University Press, 1997).
5. On power, see the wise words in Harold D. Lasswell, *Politics: Who Gets What, When, How* (London: Whittlesey House, 1936), 3. The author reminds us that "[T]he study of politics is the study of influence and the influential." Political power is about influence.
6. Carl von Clausewitz, *On War,* trans. and ed. Michael Howard and Peter Paret (1832–34) (Princeton, N.J.: Princeton University Press, 1976), 566.
7. Stephen Bungay, *Most Dangerous Enemy: A History of the Battle of Britain* (London: Aurum Press, 2000), makes what I believe to be a definitive case for the organizational independence of the RAF.
8. Carl H. Builder, *The Masks of War: American Military Styles in Strategy and Analysis* (Baltimore: Johns Hopkins University Press, 1989), ch. 5; and J. C. Wylie, *Military Strategy: A General Theory of Power Control* (Annapolis, Md.: Naval Institute Press, 1989), are both distinguished analyses of service cultures and strategic preferences that have lasting merit.
9. See Colin S. Gray, *Perspectives on Strategy* (Oxford: Oxford University Press, 2013).
10. Royal Air Force, *British Air and Space Power Doctrine,* AP3000, 4th ed. (London: Air Staff, Ministry of Defence, 2009), 16–17.
11. Richard P. Hallion, "The Future of Airpower," in *The War in the Air, 1914–1994,* ed. Alan Stephens, 382–83 (Maxwell AFB, Ala.: Air University Press, January 2001). Hallion's itemization was borrowed from Air Force Manual (AFM) 1-1, *Basic Aerospace Doctrine of the United States Air Force* (1992). For a modest check on progress, see AFDD-1, *Air Force Basic Doctrine* (2003), especially 27–33.
12. Dr. Philip Meilinger's presentation to the sixth annual conference of the Fisher Institute for Air and Space Strategic Studies, Herzliyah, Israel, May 10–11, 2010.
13. See Colin S. Gray, *The Strategy Bridge: Theory for Practice* (Oxford: Oxford University Press, 2010), especially chs. 1–2.
14. Wylie, *Military Strategy,* 72.
15. See Julian S. Corbett, *Some Principles of Maritime Strategy* (1911) (Annapolis, Md.: Naval Institute Press, 1988), part 2, ch. 1.
16. Alfred Thayer Mahan, *The Influence of Sea Power upon History, 1660–1783* (1890) (London: Methuen, 1965), 25.
17. Barry R. Posen, "Command of the Commons: The Military Foundation of U.S. Hegemony," *International Security* 28, no. 1 (Summer 2003): 5–46.

18. Clausewitz, *On War,* 75.
19. David E. Johnson, *Learning Large Lessons: The Evolving Roles of Ground Power and Air Power in the Post–Cold War Era,* MG–405–AF (Santa Monica, Calif.: RAND, 2005), is outstanding.
20. Robert L. Eichelberger, *Our Jungle Road to Tokyo* (New York: Viking, 1950).
21. Wylie, *Military Strategy,* ch. 5.
22. See Michael I. Handel, *Masters of War: Classical Strategic Thought,* 3rd ed. (London: Frank Cass, 2001), 353–60.
23. Richard P. Hallion, "Air and Space Power," in *A History of Air Warfare,* ed. John Andreas Olsen, 372 (Washington, D.C.: Potomac Books, 2010).
24. See Giulio Douhet, *The Command of the Air* (1921, 1942) (New York: Amo Press, 1972), 50.
25. See Gray, *Perspectives on Strategy,* ch. 6.
26. William Mitchell, *Winged Defense: The Development and Possibilities of Modern Air Power—Economic and Military* (1925) (New York: Dover Publications, 1988), xii, 3–4.
27. This claim flatly contradicts Robert A. Pape, *Bombing to Win: Air Power and Coercion* (Ithaca, N.Y.: Cornell University Press, 1996), though see his later assessment in "The True Worth of Air Power," *Foreign Affairs* 83, no. 2 (March/April 2004): 116–30. The great debate still lives, judging by Richard Overy, *The Bombing War: Europe, 1939–1945* (London: Allen Lane, 2013).
28. Carl H. Builder, *The Icarus Syndrome: The Role of Air Power Theory in the Evolution and Fate of the US Air Force* (New Brunswick, N.J.: Transaction Publishers 1994), is a minor classic fueled by recognition of this malady.
29. See Bungay, *The Most Dangerous Enemy,* 379, where he argues that "the Germans were out-generaled" in the Battle of Britain.
30. See James R. Corum, "Airpower Thought in Continental Europe between the Wars," in *The Paths of Heaven: The Evolution of Airpower Theory,* ed. Philip S. Meilinger, 151–81 (Maxwell AFB, Ala.: Air University Press, 1997); Scott W. Palmer, "Peasants into Pilots: Soviet Air-Mindedness as an Ideology of Dominance," *Technology and Culture* 41, no. 1 (January 2000): 1–26; and Richard Overy, *The Morbid Age: Britain and the Crisis of Civilization* (London: Penguin Books, 2010), for the dark side of the modernity brought by scientific progress. Also, Tami Davis Biddle notices that among Anglo-American air-power theorists in the 1930s "there was . . . too great a readiness to focus on the future without rigorously considering the past." She proceeds expansively to claim that "this is an endemic problem in air forces, which derive their institutional identity around claims to see and understand the future more clearly than other services do." *Rhetoric and Reality in Air Warfare: The Evolution of*

British and American Ideas about Strategic Bombing, 1914–1945 (Princeton, N.J.: Princeton University Press, 2002), 291–92.

31. The USAF waged political battle energetically but in vain to save its follow-on (to the B–52) heavy long-range manned bomber, the Mach-3 B-70, from cancellation in favor of the ICBM. See Peter J. Roman, "Strategic Bombers over the Missile Horizon, 1957–1963," *The Journal of Strategic Studies* 18, no. 1 (March 1995): 198–236.

32. Karl Marx, "The Eighteenth Brumaire of Louis Napoleon, 1852," in Marx and Friedrich Engels, *Selected Works in Two Volumes*, vol. 1 (Moscow: Foreign Languages Publishing House, 1962), 247.

33. See Robin Higham and Stephen J. Harris, eds., *Why Air Forces Fail: The Anatomy of Defeat* (Lexington: University Press of Kentucky, 2006).

34. Charalampos Nikiforidis, "A Comparative Analysis of the Prospects for Air Power and Air Forces, 1907–2010" (MA thesis, University of Reading, UK, 2010). The author helpfully identifies, contrasts, and evaluates the historical strategic performance of different models of airpower organization.

Selected Bibliography

Books

Alberts, David. *Information Age Transformation, Getting to a 21st Century Military.* Washington, D.C.: U.S. Department of Defense, CCRP Publications, 1996.

Alberts, David S., John J. Gartska, and Frederick P. Stein. *Network Centric Warfare.* Washington, D.C.: U.S. Department of Defense, CCRP Publications, 1999.

Amersfoort, Herman, Desiree Verweij, and Joseph Soeters, eds. *Moral Responsibility and Military Effectiveness.* Amsterdam: Asser Press, 2013.

Anderson, U. S. *Success-Cybernetics: The Practical Application of Human-Cybernetics.* West Nyack, N.J.: Parker, 1970.

Atkinson, Rick. *Crusade: The Untold Story of the Persian Gulf War.* New York: Houghton Mifflin Co., 1993.

Bacevich, Andrew J. *Washington Rules: America's Path to Permanent War.* New York: Metropolitan Books, 2010.

Barber, Noel. *The War of the Running Dogs: How Malaya Defeated the Communist Guerrillas 1948–60.* London: Cassel, 2004.

Barrow, John. *The Artful Universe.* New York: Little, Brown & Company, 1996.

———. *Pi in the Sky, Counting, Thinking and Being.* New York: Back Bay Books, 1993.

———. *Theories of Everything.* Oxford: Oxford University Press, 1991.

Bateson, Gregory. *Mind and Nature: A Necessary Unity.* Cresskill, N.J.: Hampton Press, 2002.

Bertalanffy, Ludwig von. *General Systems Theory.* New York: George Braziller, 1968.

Best, Steven, and Douglas Kellner. *The Postmodern Turn.* New York: Guilford Press, 1997.

Biddle, Tami Davis. *Rhetoric and Reality in Air Warfare: The Evolution of British and American Ideas about Strategic Bombing, 1914–1945.* Princeton, N.J.: Princeton University Press, 2002.

Bobbitt, Philip. *The Shield of Achilles: War, Peace, and the Course of History.* New York: Alfred A. Knopf, 2002.

Bousquet, Antoine. *The Scientific Way of War.* New York: Columbia University Press, 2009.

Boyne, Walter J. *Operation Iraqi Freedom.* New York: Tom Doherty Associates, 2003.

Brown, Shona L., and Kathleen M. Eisenhardt. *Competing on the Edge, Strategy as Structured Chaos.* Cambridge, Mass.: Harvard Business School Press, 1998.

Builder, Carl H. *The Icarus Syndrome: The Role of Air Power Theory in the Evolution and Fate of the U.S. Air Force.* New Brunswick, N.J.: Transaction Publishers 1994.

———. *The Masks of War: American Military Styles in Strategy and Analysis.* Baltimore: Johns Hopkins University Press, 1989.

Bungay, Stephen. *Most Dangerous Enemy: A History of the Battle of Britain.* London: Aurum Press, 2000.

Burton, James G. *The Pentagon Wars.* Annapolis, Md.: Naval Institute Press, 1993.

Capra, Fritjof. *The Tao of Physics.* Boston: Shambala, 1976.

———. *Turning Point: Science, Society and the Rising Culture.* New York: Bantam Books, 1982.

———. *The Web of Life: A New Scientific Understanding of Living Systems.* New York: Anchor Books, 1997.

Chaliand, Gérard, ed. *The Art of War in World History.* Berkeley: University of California Press, 1994.

Cimbala, Stephen J., and Keith A. Dunn, eds. *Conflict Termination and Military Strategy: Coercion, Persuasion, and War.* Boulder, Colo.: Westview Press, 1993.

Clausewitz, Carl von. *On War.* Translated and edited by Michael Howard and Peter Paret. Princeton, N.J.: Princeton University Press, 1976.

———. Carl von. *On War.* Edited by Anatol Rapoport. Harmondsworth: Penguin, 1982.

Clodfelter, Mark. *Beneficial Bombing: The Progressive Foundations of American Airpower, 1917–1945.* Lincoln: University of Nebraska Press, 2010.

Cohen, Eliot A. *Supreme Command.* New York: The Free Press, 2002.

Coker, Christopher. *Humane Warfare.* London: Routledge, 2002.

———. *War in the Age of Risk.* Cambridge: Polity Press, 2009.

Conant, James Bryant. *Two Modes of Thought.* New York: Trident Press, 1964.

Coram, Robert. *Boyd: The Fighter Pilot Who Changed the Art of War.* Boston: Little, Brown, 2002.

Corbett, Julian S. *Some Principles of Maritime Strategy.* (1911). Annapolis, Md.: Naval Institute Press, 1988.

Cordesman, Anthony. *The Iraq War: Strategy, Tactics, and Military Lessons.* Westport, Conn.: Praeger, 2003.

Corum, James, and Wray Johnson. *Airpower in Small Wars: Fighting Insurgents and Terrorists.* Lawrence: University Press of Kansas, 2003.

Coveney, Peter, and Roger Highfield. *The Arrow of Time.* London: Flamingo, 1991.

———. *Frontiers of Complexity.* New York: Ballantine Books, 1996.

Cox, Sebastian, and Peter Gray, eds. *Air Power History: Turning Points from Kitty Hawk to Kosovo.* London: Frank Cass, 2002.

Creveld, Martin van. *The Age of Airpower.* New York: Public Affairs, 2011.

———. *Command in War.* Cambridge, Mass.: Harvard University Press, 1987.

Davis, Richard G. *Bombing the European Axis Powers: A Historical Digest of the Combined Bomber Offensive, 1939–1945*. Maxwell AFB, Ala.: Air University Press, 2006.

Dawkins, Richard. *The Blind Watchmaker*. New York: W. W. Norton & Company, 1986.

———. *The Selfish Gene*. Oxford: Oxford University Press, 1976.

De Saxe, Maurice. *Reveries on the Art of War*. Translated and edited by Brig. Gen. Thomas R. Phillips. Mineola, N.Y.: Dover Publications, Inc., 2007.

DoD Report to Congress on NCW. Washington, D.C.: U.S. Department of Defense, CCRP Publications, 2001.

Douhet, Giulio. *The Command of the Air*. (1921, 1942). New York: Amo Press, 1972.

Echevarria II, Antulio J. *Clausewitz and Contemporary War*. Oxford: Oxford University Press, 2007.

Ehrhard, Thomas P. *Making the Connection: An Air Strategy Analysis Framework*. Maxwell AFB, Ala.: Air University Press, 1995.

Eichelberger, Robert L. *Our Jungle Road to Tokyo*. New York: Viking, 1950.

Eve, Raymond A., et al. *Chaos, Complexity, and Sociology*. London: Sage, 1997.

Eysturlid, Lee, and John Coetzee, eds. *Philosophers of War*. Santa Barbara, Calif.: Praeger, 2013.

Fadok, David S. *John Boyd and John Warden: Air Power's Quest for Strategic Paralysis*. Maxwell AFB, Ala.: Air University Press, 1995.

Ferguson, Niall. *Civilization: The West and the Rest*. London: Allen Lane, 2011.

Fitzgibbon, Constantine. *The Winter of the Bombs*. New York: Norton, 1957.

Freedman, Lawrence. *Strategy: A History*. Oxford: Oxford University Press, 2013.

———, ed. *War*. Oxford: Oxford University Press, 1994.

Fuller, J. F. C. *The Foundations of the Science of War*. Fort Leavenworth, Kan.: Combat Studies Institute Press, 2012.

———. *A Military History of the Western World*. 3 vols. New York: Funk and Wagnalls, 1954–1957; reprint, New York: Da Capo, 1987.

Gardner, Howard. *The Mind's New Science: A History of the Cognitive Revolution*. New York: Basic Books, 1985.

Gaston, James C. *Planning the American Air War: Four Men and Nine Days in 1941*. Washington, D.C.: National Defense University Press, 1982.

Gat, Azar. *The Development of Military Thought: The Nineteenth Century*. Oxford: Clarendon Press, 1992.

———. *A History of Military Thought: From the Enlightenment to the Cold War*. Oxford: Oxford University Press, 2001.

———. *The Origins of Military Thought: From the Enlightenment to Clausewitz*. Oxford: Clarendon Press, 1989.

Gell-Mann, Murray. *The Quark and the Jaguar*. New York: Freeman & Company, 1994.

George, F. H. *Cybernetics.* London: Hodder and Stoughton, 1971.

Glad, Betty, ed. *Psychological Dimensions of War.* Newbury Park, Calif.: Sage Press, 1990.

Glaser, Charles L. *Analyzing Strategic Nuclear Policy.* Princeton, N.J.: Princeton University Press, 1990.

Gleick, James. *Chaos.* New York: Penguin Books, 1987.

Gompert, David, Richard Kugler, and Martin Libicki. *Mind the Gap: Promoting a Transatlantic Revolution in Military Affairs.* Washington D.C.: National Defense University Press, 1997.

Gordon, Michael, and Bernard Trainor. *Cobra II: The Inside Story of the Invasion and Occupation of Iraq.* New York: Random House, 2006.

———. *The Generals' War.* Boston: Little, Brown and Company, 1995.

Goss, Hilton P. *Civilian Morale Under Aerial Bombardment 1914–1939.* 2 vols. Maxwell AFB, Ala.: Documentary Research Division, Air University Libraries, Air University, 1948.

Gray, Colin S. *Airpower for Strategic Effect.* Maxwell AFB, Ala.: Air University Press, 2012.

———. *Modern Strategy.* Oxford: Oxford University Press, 1999.

———. *Perspectives on Strategy.* Oxford: Oxford University Press, 2013.

———. *The Strategy Bridge: Theory for Practice.* Oxford: Oxford University Press, 2010.

Hagan, Kenneth J., and Ian J. Bickerton. *Unintended Consequences: The United States at War.* London: Reaktion Books, 2007.

Halberstam, David. *War in a Time of Peace.* London: Bloomsbury, 2001.

Hall, Edward. *Beyond Culture.* New York: Anchor Books, 1976.

Hall, R. Cargill, ed. *Case Studies in Strategic Bombardment.* Washington, D.C.: U.S. Government Printing Office, 1998.

Hallion, Richard P., ed. *Air Power Confronts An Unstable World.* London: Brassey's, 1997.

———. *Storm over Iraq: Air Power and the Gulf War.* Washington, D.C.: Smithsonian Institution Press, 1992.

Hammond, Grant T. *The Mind of War: John Boyd and American Security.* Washington, D.C.: Smithsonian Institution Press, 2001.

Handel, Michael I. *Masters of War: Classical Strategic Thought,* 3rd ed. London: Frank Cass, 2001.

Harrison, Lawrence E., and Samuel P. Huntington. *Culture Matters.* New York: Basic Books, 2000.

Hawking, Stephen. *A Brief History of Time.* New York: Bantam Books, 1988.

Higham, Robin, and Stephen J. Harris, eds. *Why Air Forces Fail: The Anatomy of Defeat.* Lexington: University Press of Kentucky, 2006.

Hirshleifer, Jack. *Disaster and Recovery: A Historical Survey.* Santa Monica, Calif.: RAND Corporation Research Memorandum 3079 (RM-3079), 1963.

Hosmer, Stephen T. *Psychological Effects of U.S. Air Operations in Four Wars, 1941–1991.* Santa Monica, Calif.: RAND Corporation, 1996.

Howard, Michael. *Causes of War and Other Essays.* London: Counterpoint, 1983.

———. *Clausewitz.* Oxford: Oxford University Press, 1983.

Ienaga, Saburo. *The Pacific War 1931–1945.* New York: Pantheon, 1978.

Ignatieff, Michael. *Virtual War: Kosovo and Beyond.* London: Chatto and Windus, 2000.

Ikle, Charles. *The Social Impact of Bomb Destruction.* Norman: University of Oklahoma Press, 1958.

Janis, Irving L. *Air War and Emotional Stress.* Westport, Conn.: Greenwood Press, 1951.

Johnson, David E. *Learning Large Lessons: The Evolving Roles of Ground Power and Air Power in the Post–Cold War Era,* MG–405–AF. Santa Monica, Calif.: RAND, 2005.

Joint Chiefs of Staff. *Joint Vision 2010.* Washington, D.C.: U.S. Department of Defense, 1997.

Karlins, Marvin, and Lewis Andrews. *Biofeedback: Turning on the Power of Your Mind.* Philadelphia: J. B. Lippincott Co, 1973.

Karnow, Stanley. *Vietnam: A History.* New York: Penguin Books, 1984.

Kaysen, Carl. *Note on Some Historic Principles of Target Selection.* Santa Monica, Calif.: RAND Corporation, Project RAND Research Memorandum 189 (RM-189), July 15, 1949.

Keaney, Thomas A., and Eliot A Cohen. *A Revolution in Warfare: Air Power in the Persian Gulf.* Annapolis, Md.: Naval Institute Press, 1995.

Keaney, Thomas, and Thomas Mahnken, eds. *War in Iraq: Planning and Execution.* London: Routledge, 2007.

Kellert, Stephen. *In the Wake of Chaos.* Chicago: University of Chicago Press, 1993.

Kelly, Susanne, and May Ann Allison. *The Complexity Advantage: How the Sciences Can Help Your Business Achieve Peak Performance.* New York: McGraw-Hill, 1999.

Krepinevich, Andrew F. *The Army and Vietnam.* Baltimore, Md.: Johns Hopkins University Press, 1986.

Kuhn, Thomas S. *The Structure of Scientific Revolutions.* Chicago: Chicago University Press, 1970.

Lambeth, Benjamin S. *NATO's Air War for Kosovo.* Santa Monica, Calif.: RAND, 2001.

———. *The Transformation of American Air Power.* Ithaca, N.Y.: Cornell University Press, 2000.

———. *The Unseen War: Allied Air Power and the Takedown of Saddam Hussein.* Annapolis, Md.: Naval Institute Press, 2013.

Lasswell, Harold D. *Politics: Who Gets What, When, How.* London: Whittlesey House, 1936.

Liddell Hart, B. H. *Strategy.* New York: Meridian, 1991.
Lindley-French, Julian, and Yves Boyer, eds. *The Oxford Handbook of War.* Oxford: Oxford University Press, 2012.
Linn, Brian McAllister. *The Echo of Battle: The Army's Way of War.* Cambridge, Mass.: Harvard University Press, 2007.
Lock-Pullan, Richard. *US Intervention Policy and Army Intervention.* London: Routledge, 2006.
Luttwak, Edward N. *The Grand Strategy of the Byzantine Empire.* Cambridge, Mass.: The Belknap Press of the Harvard University Press, 2009.
Mahan, Alfred Thayer. *The Influence of Sea Power upon History, 1660–1783.* (1890). London: Methuen, 1965.
Maltz, Maxwell. *Psycho-Cybernetics.* Hollywood, Calif.: Wilshire Book Company, 1971.
Marx, Karl, and Friedrich Engels. *Selected Works in Two Volumes,* vol. 1. Moscow: Foreign Languages Publishing House, 1962.
Mason, Richard A. *Air Power: A Centennial Appraisal.* London: Brassey's, 1994.
May, Ernest. *Lessons of the Past.* New York: Oxford University Press, 1976.
McInnes, Colin. *Spectator-Sport War.* London: Rienner, 2002.
McKeown, Max. *Adaptability: The Art of Winning in an Age of Uncertainty.* London: Kogan Page, 2012.
Meilinger, Philip S., *Ten Propositions Regarding Air Power.* Washington, D.C.: Air Force History and Museums Program, 1995.
———, ed. *The Paths of Heaven: The Evolution of Airpower Theory.* Maxwell AFB, Ala.: Air University Press, December 1997.
Merry, Uri. *Coping with Uncertainty: Insights from the New Sciences of Chaos, Self Organization, and Complexity.* Westport, Conn.: Praeger, 1995.
Miller, David. *A Pocket Popper.* Oxford: Fontana, 1983.
Minsky, Marvin. *Society of the Mind.* New York: Simon & Schuster, 1988.
Mitchell, William. *Winged Defense: The Development and Possibilities of Modern Air Power—Economic and Military.* (1925). New York: Dover Publications, 1988.
Momyer, William M. *Air Power in Three Wars.* Washington, D.C.: U.S. Government Printing Office, 1978.
Morgan, Gareth. *Images of Organizations.* New York: Sage, 1986.
Mueller, Karl, ed. *Precision and Purpose: Airpower in the Libyan Civil War.* Santa Monica, Calif.: RAND, 2014.
Murray, Williamson, and Robert Scales. *The Iraqi War: A Military History.* Cambridge, Mass.: Harvard University Press, 2003.
Naylor, Sean. *Not a Good Day to Die: The Untold Story of Operation Anaconda.* New York: Berkley Caliber, 2005.
Neumann, John von. *The Computer and the Brain.* New Haven, Conn.: Yale University Press, 1958.
Olsen, John Andreas, ed. *Air Commanders.* Washington, D.C.: Potomac Books, 2012.

———. *Asymmetric Warfare.* Oslo: Royal Norwegian Air Force Academy Press, 2002.

———, ed. *Global Air Power.* Washington, D.C.: Potomac Books, 2011.

———, ed. *A History of Air Warfare.* Washington, D.C.: Potomac Books, 2010.

———. *John Warden and the Renaissance of American Air Power.* Washington, D.C.: Potomac Books, 2007.

Ornstein, Robert. *The Evolution of Consciousness.* New York: Prentice-Hall, 1991.

Osinga, Frans. *Science, Strategy and War: The Strategic Theory of John Boyd.* London: Routledge, 2007.

Overy, Richard. *The Air War 1939–1945.* London: Papermac, 1987.

———. *The Bombing War: Europe, 1939–1945.* London: Allen Lane, 2013.

———. *The Bombing War: Europe 1939–1945.* New York: Penguin Books, 2013.

———. *The Morbid Age: Britain and the Crisis of Civilization.* London: Penguin Books, 2010.

Owen, Robert C., ed. *Deliberate Force: A Case Study in Effective Air Campaigning.* Maxwell AFB, Ala.: Air University Press, 2000.

Pape, Robert A. *Bombing to Win: Air Power and Coercion.* Ithaca, N.Y.: Cornell University Press, 1996.

Paret, Peter. *Clausewitz and the State: The Man, His Theories, and His Times.* Princeton, N.J.: Princeton University Press, 1976.

Paret, Peter, with Gordon Craig and Felix Gilbert, eds. *Makers of Modern Strategy.* Princeton, N.J.: Princeton University Press, 1986.

Pearce, Chilton. *The Crack in the Cosmic Egg: Challenging Constructs of Mind and Reality.* New York: Pocket Books, 1974.

Pellegrini, Robert P. *The Links Between Science and Philosophy and Military Theory: Understanding the Past; Implications for the Future.* Maxwell AFB, Ala.: Air University Press, 1995.

Penrose, Roger. *The Emperor's New Mind: Concerning Computers, Minds and the Laws of Physics.* Oxford: Oxford University Press, 1989.

———. *Shadows of the Mind: A Search for the Missing Science of Consciousness.* Oxford: Oxford University Press, 1994.

Pentland, Pat. *Center of Gravity Analysis and Chaos Theory.* Maxwell AFB, Ala.: Air University Press, April 1993.

Perrow, Charles. *Complex Organizations: A Critical Essay.* Glenview, Ill.: Scott, Foresman, 1972.

———. *Normal Accidents.* Princeton, N.J.: Princeton University Press, 1999.

Polanyi, Michael. *Knowing and Being.* London: Routledge and Kegan Paul, 1969.

Prigogine, Ilya, and Isabelle Stengers, *Order Out of Chaos.* London: Bantam Books, 1984.

Rasmussen, Mikkel. *The Risk Society at War.* Cambridge: Cambridge University Press, 2007.

Reynolds, Richard T. *Heart of the Storm: The Genesis of the Air Campaign Against Iraq* (Maxwell AFB, Ala.: Air University, 1995).

Ricks, Thomas E. *Fiasco: The American Military Adventure in Iraq*. London: Penguin Books, 2007.

Ritzer, George, ed. *The Blackwell Companion to Major Contemporary Social Theorists*. Oxford: Blackwell Publishing, 2003.

Rose, Stephen. *The Making of Memory*. London: Bantam Press, 1992.

Ruelle, David. *Chance and Chaos*. Princeton, N.J.: Princeton University Press, 1991.

Rumpf, Hans. *The Bombing of Germany*. Translated by Edward Fitzgerald. New York: Holt, Rinehart and Winston, 1963.

Ryle, Gilbert. *The Concept of Mind*. London: Hutchinson, 1949.

Schelling, Thomas C. *Arms and Influence*. New Haven, Conn.: Yale University Press, 1966.

Shaw, Martin. *The New Western Way of War*. Cambridge: Polity Press, 2005.

Sheehan, Neil. *A Bright Shining Lie*. London: Jonathan Cape, 1989.

Sigal, Leon. *Fighting to a Finish: The Politics of War Termination in the United States and Japan, 1945*. Ithaca, N.Y.: Cornell University Press, 1988.

Simpson, Emile. *War from the Ground Up: Twenty-first Century Combat as Politics*. London: C. Hurst & Co., 2012.

Smith, Phil A. *Bombing to Surrender: The Contributions of Airpower to the Collapse of Italy, 1943*. Maxwell AFB, Ala.: Air University Press, 1998.

Stephens, Alan, ed. *The War in the Air: 1914–1994*. Fairbairn, Australia: Air Power Studies Centre, 1994; reprint Maxwell AFB, Ala.: Air University, 2001.

Stephens, Alan, and Nicola Baker. *Making Sense of War: Strategy for the 21st Century*. Melbourne: Cambridge University Press, 2006.

Titmuss, Richard M. *Problems of Social Policy: History of the Second World War, United Kingdom Civil Series*. London: Her Majesty's Stationery Office, 1950.

Titus, James. *The Battle of Khafji: An Overview and Preliminary Analysis*. Maxwell AFB, Ala.: Air University Press, 1996.

Tzu, Sun. *The Art of War*. Edited by Samuel B. Griffith. Oxford: Oxford University Press, 1963.

———. *The Art of War*. Translated from the Chinese by Lionel Giles, MA, 1910. Hollywood, Fla.: Simon & Brown, 2010.

U.S. Army and U.S. Marine Corps, Field Manual 3-24 and Marine Corps Warfighting Publication 3-33.5. *Counterinsurgency*. Chicago: University of Chicago Press, 2007.

United States Strategic Bombing Survey. *The Effects of Strategic Bombing on German Morale*, vol. I. Washington, D.C.: Government Printing Office, 1947.

———. *Summary Report (European War)*. Washington, D.C.: United States Government Printing Office, 1946.

Waldrop, Mitchell. *Complexity: The Emerging Science at the Edge of Order and Chaos*. London: Viking, 1993.

Waltz, Kenneth. *Theory of International Politics*. Reading, Mass.: Addison-Wesley Publishing Co., 1979.

Warden, John A. *The Air Campaign.* Washington, D.C.: Brassey's, 1989.
Watson, Peter. *A Terrible Beauty.* London: Phoenix Press, 2000.
Watts, Barry. *The Foundations of US Air Doctrine, the Problem of Friction in War.* Maxwell AFB, Ala.: Air University Press, 1984.
Weigley, Russell F. *The American Way of War: A History of United States Military Strategy and Policy.* Bloomington: Indiana University Press, 1977.
White, Charles Edward. *The Enlightened Soldier: Scharnhorst and the Militärische Gesellschaft in Berlin, 1801–1805.* Westport, Conn.: Praeger Publishers, 1989.
Wiener, Norbert. *The Human Use of Human Beings: Cybernetics and Society.* Cambridge, Mass.: MIT Press, 1950.
———. *The Human Use of Human Beings: Cybernetics and Society.* New York: Avon Books, 1967.
Williams, Kevin E. *In Search of the Missing Link: Relating Destruction to Outcome in Airpower Applications.* Maxwell AFB, Ala.: Air University Press, 1994.
Wilson, E. O. *On Human Nature.* Cambridge, Mass.: Harvard University Press, 1978.
Woodward, Bob. *Obama's Wars.* New York: Simon & Schuster, 2010.
———. *Plan of Attack.* New York: Simon & Schuster, 2004.
Wylie, J. C. *Military Strategy: A General Theory of Power Control.* Annapolis, Md.: Naval Institute Press, 1989.
Yarnold, K. W. *Lessons on Morale to be Drawn from Effects of Strategic Bombing on Germany: With Special Reference to Psychological Warfare.* Technical Memorandum, ORO-T-2. Washington, D.C.: Operations Research Office, 1949.
Yule, Henry, ed. *The Book of Ser Marco Polo, the Venetian, Concerning the Kingdom and Marvels of the East.* London: John Murray, 1903.
Zolli, Andrew, and Ann Marie Healy. *Resilience: Why Things Bounce Back.* New York: Simon & Schuster, 2012.

ARTICLES, REPORTS, AND THESES

Abizaid, John. "Combined Civil/Military Responses to National and International Events." The Future Australian Defence Force: Learning from the Past, Planning for the Future. Canberra: Australian Defence College and Royal United Services Institute Seminar 2007, May 16, 2007.
Bacevich, Andrew J. "Surge to Nowhere." *Washington Post,* January 20, 2008.
Barabási, Albert-László, and Eric Bonabeau, "Scale-Free Networks." *Scientific American,* May 2003.
Beinhocker, Eric D. "Robust Adaptive Strategies." *Sloan Management Review* (Spring 1999).
Beyerchen, Alan. "Clausewitz, Nonlinearity, and the Unpredictability of War." *International Security* 17, no. 3 (Winter 1992).

Biddle, Stephen. "Afghanistan and the Future of Warfare." *Foreign Affairs* 82 (March/April 2003).

Boot, Max. "Beyond the 3-Block War." *Armed Forces Journal* (March 2006).

———. "The New American Way of War." *Foreign Affairs* 82 (July/August 2003).

Budiansky, Stephen. "Of Tools and Tasks: Air War—Striking in Ways We Haven't Seen." *Washington Post*, April 26, 2003.

Byman, Daniel L., and Matthew C. Waxman. "Kosovo and the Great Air Power Debate." *International Security* 24, no. 4 (Spring 2000).

Caproni, Gianni. "Memorandum on 'Air War.'" Air Force History Research Agency, file no. 168.661-129, 1917.

Church, Michael. "Organizing Simply for Complexity: Beyond Metaphor Towards Theory." *Long Range Planning* 32, no. 4 (1999).

Clark, Christopher. "The First Calamity." *The London Review of Books*, August 29, 2013.

Cockburn, Patrick. "What We Haven't Heard About Iraq." *London Review of Books*, October 10, 2013.

Coleman, Henry. "What Enables Self-Organizing Behavior in Businesses." *Emergence* 1, iss. 1 (1999).

Echevarria, Antulio J. "Moltke and the German Military Tradition: His Theories and Legacies." *Parameters* XXVI (Spring 1996).

Eisenhardt, Kathleen, and Donald N. Sull. "Strategy as Simple Rules." *Harvard Business Review* (January 2001).

Fabyanic, Thomas A. "War Doctrine, and the Air War College—Some Implications for the U.S. Air Force." *Air University Review* XXXVII (January–February 1986).

Fracker, Martin L. "Psychological Effects of Aerial Bombardment." *Airpower Journal* VI (Fall 1992).

Gaddis, John Lewis. "International Relations Theory and the End of the Cold War." *International Security* 17 (Winter 1992/93).

Gardner, Nikolas. "Resurrecting the 'Icon': The Enduring Relevance of Clausewitz's *On War*." *Strategic Studies Quarterly* (Spring 2009).

George, Alexander. "Emotional Stress and Air War—A Lecture Given at the Air War College." Air Force Historical Research Agency, file no. K239.716251-65, November 1951.

Grossman, Elaine M. "The Halt Phase Hits a Bump." *Air Force Magazine* (April 2001).

Gventer, Celeste Ward. "A False Promise of 'Counterinsurgency.'" *New York Times*, December 1, 2009.

Harvey, Frank. "Getting NATO's Success in Kosovo Right: The Theory and Logic of Counter-Coercion." *Conflict Management and Peace Science* 23 (2006).

Hasik, James. "Beyond the Briefing: Theoretical and Practical Problems in the Works and Legacy of John Boyd." *Contemporary Security Policy* 34, no. 3 (December 2013).

Hirschman, Albert O. "The Search for Paradigms as a Hindrance to Understanding." *World Politics* 22 (April 1970).
Howard, Michael. "Narratives of War." *The Times Literary Supplement*, April 3, 2013.
Hughes, Thomas. "The Cult of the Quick." *Air Power Journal* XV, no. 4 (Winter 2001).
Huntington, Samuel P. "The Clash of Civilisations?" *Foreign Affairs* (Summer 1993).
International Institute for Strategic Studies. *Strategic Survey 2001–2002*. London: IISS, 2002.
Johnston, Edward S. "A Science of War." *The Command and General Staff Quarterly* XIV (June 1934).
Keegan, John. "Please Mr Blair, Never Take Such A Risk Again." *Daily Telegraph* (London), June 6, 1999.
Kober, Avi. "Nomology vs Historicism: Formative Factors in Modern Military Thought." *Defense Analysis* 10, no. 3 (1994).
Krulak, Charles C. "Obituary." *Inside the Pentagon*, March 13, 1997.
———. "The Strategic Corporal: Leadership in the Three Block War." *Marines Magazine* (January 1999).
Kubik, Timothy R. W. "Is Machiavelli's Canon Spiked? Practical Readings in Military History." *The Journal of Military History* 61 (January 1997).
Lambert, A. P. N. "The Psychological Impact of Air Power Based on Case Studies since the 1940s." MA thesis, Cambridge University, 1995.
Lambeth, Benjamin S. "Pitfalls in Force Planning: Structuring America's Tactical Air Arm." *International Security* 10 (Fall 1985).
———. "The Technology Revolution in Air Warfare." *Survival* 39, no. 1 (Spring 1997).
Latham, Andrew. "Warfare Transformed: A Braudelian Perspective on the 'Revolution in Military Affairs.'" *European Journal of International Relations* 8, no. 2 (2002).
Link, Charles. "21st Century Armed Forces—Joint Vision 2010." *Joint Forces Quarterly* (Autumn 1996).
Lissack, Michael. "Complexity: The Science, its Vocabulary, and its Relation to Organizations." *Emergence* 1, iss. 1 (1999).
Mann, Steven. "Chaos Theory and Strategic Thought." *Parameters* XXII, no. 2 (Autumn 1992).
Marion, Russ, and Josh Bacon. "Organizational Extinction and Complex Systems." *Emergence* 1, no. 4 (1999).
Meilinger, Phillip S. "Busting the Icon: Restoring Balance to the Influence of Clausewitz." *Strategic Studies Quarterly* (Fall 2007).
Mueller, Karl. "Strategies of Coercion: Denial, Punishment and the Future of Air Power." *Security Studies* 7, no. 2 (Winter 1997/98).
Nikiforidis, Charalampos. "A Comparative Analysis of the Prospects for Air Power and Air Forces, 1907–2010." MA thesis, University of Reading, UK, 2010.
O'Hanlon, Michael. "A Flawed Masterpiece." *Foreign Affairs* (May/June 2002).

O'Neill, Robert. "Restoring Utility to Armed Force in the 21st Century." Strategic and Defence Studies Centre 40th Anniversary Seminar Series. Canberra: Australian National University, August 15, 2006.

Osinga, Frans. "'Getting' a Discourse on Winning and Losing: A Primer on Boyd's Theory of Intellectual Evolution." *Contemporary Security Policy* 34, no. 3 (December 2013).

Palmer, Scott W. "Peasants into Pilots: Soviet Air-Mindedness as an Ideology of Dominance." *Technology and Culture* 41, no. 1 (January 2000).

Pape, Robert A. "Coercion and Military Strategy: Why Denial Works and Punishment Doesn't." *The Journal of Strategic Studies* 15 (December 1992).

———. "The True Worth of Air Power." *Foreign Affairs* 83, no. 2 (March/April 2004).

———. "Why Japan Surrendered." *International Security* 18 (Fall 1993).

Partlow, Joshua. "Karzai Wants US to Reduce Military Operations in Afghanistan." *Washington Post,* November 14, 2010.

Pentland, Pat A. "Theatre Strategy Development." Unpublished manuscript, 1993–94.

Posen, Barry R. "Command of the Commons: The Military Foundation of U.S. Hegemony." *International Security* 28, no. 1 (Summer 2003).

———. "The War for Kosovo: Serbia's Political-Military Strategy." *International Security* 24, no. 4 (Spring 2000).

Press, Daryl G. "The Myth of Air Power in the Persian Gulf War and the Future of War." *International Security* 26, no. 2 (Fall 2001).

Quester, George H. "The Impact of Strategic Air Warfare." *Armed Forces and Society* 4 (February 1978).

Riper, Paul van, and Robert H. Scales Jr., "Preparing for War in the 21st Century." *Parameters* (Autumn 1997).

Roman, Peter J. "Strategic Bombers over the Missile Horizon, 1957–1963." *The Journal of Strategic Studies* 18, no. 1 (March 1995).

Rosinski, Herbert. "Scharnhorst to Schlieffen: The Rise and Decline of German Military Thought." *Naval War College Review* XXIX (Summer 1976).

Royal Air Force. *British Air and Space Power Doctrine,* AP3000, 4th ed. London: Air Staff, Ministry of Defence, 2009.

Salvaneschi, Nino. "Let Us Kill the War: Let us Aim at the Heart of the Enemy." Maxwell AFB, Ala.: Air Force Historical Research Agency File Number 168.661-129, 1917.

Scales, Robert L. "Checkmate by Operational Maneuver." *Armed Forces Journal International* (October 2001).

Schaffel, Kenneth. "Muir S. Fairchild: Philosopher of Air Power." *Aerospace Historian* 33 (Fall 1986).

Shanker, Thom, and Eric Schmitt. "Pentagon Weighs Strategy Change to Deter Terror." *New York Times,* July 5, 2005.

Shrader, Charles W. "The Influence of Vegetius' *De re militari.*" *Military Affairs* 45 (December 1981).

Simpson, Emile. "The Fusion of Military and Political Activity in 21st Century Combat." A presentation at the Strategic and Defence Studies Centre, Australian National University, Canberra, October 18, 2013.

Sink, Jerry T. "Coercive Air Power: The Theory of Leadership Relative Risk." Unpublished essay. Maxwell AFB, Ala.: USAF School of Advanced Airpower Studies, April 1993.

Smith, R. David. "The Inapplicability Principle: What Chaos Means for Social Science." *Behavioral Science* 40 (January 1995).

Stigler, Andrew L. "A Clear Victory for Air Power: NATO's Empty Threat to Invade Kosovo." *International Security* 27, no. 3 (Winter 2002/03).

Storr, Jim. "Neither Art Nor Science—Towards a Discipline of Warfare." *RUSI Journal* (April 2001).

The Strategic Aerospace Warfare Study Panel. "Aerospace Power for the 21st Century: A Theory to Fly By." White Paper. Maxwell AFB, Ala., October 4, 1996.

Sullivan, Gordon R., and James M. Dublik. "War in the Information Age." *Military Review* (April 1994).

Vincent, Gary. "In the Loop, Superiority in Command and Control." *Airpower Journal* VI (Summer 1992).

———. "A New Approach to Command and Control, the Cybernetic Design." *Airpower Journal* VII (Summer 1993).

Warden, John A., III. "The Enemy as a System." *Airpower Journal* IX (September 1995).

Watts, Barry D. *Clausewitzian Friction and Future War.* Institute for National Strategic Studies McNair Papers, No. 52. Washington, D.C.: National Defense University, October 1996.

Wilson, Don. "Long Range Airplane Development." Air Force Historical Research Agency, Maxwell AFB, Ala., file no. 248.211-17, November 1938.

Winton, Harold R. "An Imperfect Jewel: Military Theory and the Military Profession." *The Journal of Strategic Studies* 34, no. 6 (December 2011).

Yazhou, Liu. "Interview with Lieutenant General Liu Yazhou." *Heartland: Eurasian Review of Geopolitics.* Hong Kong: Gruppo Editoriale, L'Esspresso/Cassan Press, January 2005.

INDEX

About the Editor and Authors

John Andreas Olsen is a colonel in the Royal Norwegian Air Force currently assigned to NATO Headquarters in Brussels, a professor at the Norwegian Institute for Defence Studies, a non-resident senior fellow of the Mitchell Institute for Aerospace Studies, a fellow of the Royal Swedish Academy of War Sciences and a member of the RUSI Journal editorial board. His previous assignments include tours as dean of the Norwegian Defence University College and head of its division for strategic studies. Colonel Olsen is a graduate of the German Command and Staff College and has served both as liaison officer to the German Operational Command in Potsdam and as military assistant to the Norwegian Embassy in Berlin. He has a doctorate in history and international relations from De Montfort University, a master's degree in contemporary literature from the University of Warwick, and a master's degree in English from the University of Trondheim. Professor Olsen's books on airpower appear on the professional reading lists of the Royal Air Force, Royal Australian Air Force, Royal Netherlands Air Force, Royal Norwegian Air Force and the U.S. Air Force, and include *Strategic Air Power in Desert Storm* (2003) and *John Warden and the Renaissance of American Air Power* (2007); coauthor of *Destination NATO: Defence Reform in Bosnia and Herzegovina, 2003–2013* (2013); editor of *On New Wars* (2006), *A History of Air Warfare* (2010), *Global Air Power* (2011), *Air Commanders* (2012), and *European Air Power* (2014); and coeditor of *The Evolution of Operational Art: From Napoleon to the Present* (2011) and *The Practice of Strategy: From Alexander the Great to the Present* (2012).

Peter R. Faber currently heads the International Relations and Security Network (ISN) in Zurich, Switzerland. He was a faculty member at the U.S. National War College and also served in the Department of Defense for thirty years, where he worked on numerous policy- and planning-related issues up to the secretary of defense level. Colonel (ret.) Faber holds five advanced degrees, including two from Yale University. His areas of specialization include U.S. national security policy and planning; Western military history, theory, and strategy; European security issues; and global terrorism. Over the years he has taught at nine academic and professional schools, including the North Atlantic Treaty Organization (NATO) Defence College in Rome, the Elliott School of International Affairs (George Washington University, Washington, D.C.), the U.S. Air Force's School of Advanced Airpower Studies, and the U.S. Air Force Academy. He has written numerous articles and studies on security-related topics and has delivered lectures in twenty-four countries.

Colin S. Gray is a strategic theorist and defense analyst at the Department of International Politics and Strategic Studies at the University of Reading, England. Educated at the University of Manchester (BA, economics, 1965) and at Lincoln College, Oxford University (D.Phil., international politics, 1970), he has written pioneering and controversial studies on airpower theory, nuclear strategy, arms control, maritime strategy, and geopolitics. Dr. Gray is a member of the editorial boards of *Comparative Studies Strategy, Journal of Strategic Studies, Strategic Studies Quarterly, Naval War College Review,* and the *Journal of Terrorism and Organized Crime.* He has lectured on defense, security, and foreign affairs in Europe and North America, as well as in China, Israel, and Australia. Professor Gray's publications include *Explorations in Strategy* (1996), *Modern Strategy* (1999), *Strategy for Chaos: Revolution in Military Affairs and the Evidence of History* (2002), *The Sheriff: America's Defense and New World Order* (2004), *Another Bloody Century: Future Warfare* (2005), *Strategy and History: Essays on Theory and Practice* (2006), *War, Peace, and International Relations: An Introduction to Strategic History* (2007), *Fighting Talk: Forty Maxims on War, Peace and Strategy* (2007), *National Security Dilemmas: Challenges and Opportunities* (2009), *The Strategy Bridge: Theory and Practice* (2010), *Airpower for Strategic Effect* (2012), and *Perspectives on Strategy* (2013).

Frans P. B. Osinga is an air commodore in the Royal Netherlands Air Force and a professor of war studies, head of the Military Operational Art and Science Section, and chair of the war studies program at the Netherlands' Defence Academy. After completing a tour at the Clingendael Institute of International Relations as the Ministry of Defence's senior research fellow, he served at NATO Allied Command Transformation as the liaison officer for the newly established Joint Air Power Competence Center, and then as assistant professor in military science at the Royal Netherlands Military Academy. Trained as an F-16 pilot, Commodore Osinga graduated from the Royal Netherlands Military Academy, the advanced staff course of the Netherlands Defence College, and the School of Advanced Airpower Studies, and he holds a PhD in political science from Leiden University. His publications and lectures cover topics such as the theory and practice of airpower, European Union defense policy, military transformation, asymmetric warfare, and terrorism. He is the author of *Science, Strategy and War: The Strategic Theory of John Boyd* (2006) and coeditor of *A Transformation Gap? American Innovations and European Military Change* (2010).

Alan Stephens is a visiting fellow at the University of New South Wales (UNSW) and a member of the Sir Richard Williams Foundation. Previously he has been a lecturer at UNSW, the Royal Australian Air Force historian, an advisor to the Australian federal parliament on foreign affairs and defense, a visiting fellow at the Strategic and Defence Studies Centre at the Australian National University, and a pilot in the Royal Australian Air Force (RAAF), where his experience included the command of an operational squadron and a tour in Vietnam. Dr. Stephens has published and lectured extensively. He is the author of *Going Solo: The Royal Australian Air Force 1946–1971* (1995), *High Fliers: Leaders of the RAAF* (1996), and *The Australian Centenary History of Defence, Volume 2, The Royal Australian Air Force* (2002); coauthor of *Making Sense of War: Strategy for the 21st Century* (2006); and editor of *The War in the Air: 1914–1994* (1994) and *New Era Security* (1996). In 2008 he was made a member of the Order of Australia for his contribution to air force history and to Australian airpower strategy.

John A. Warden III is widely acknowledged as the main architect of the theory underlying the air campaign that liberated Kuwait from Iraqi occupation in 1991. After earning his fighter wings in 1966, he flew 250 combat missions in Vietnam. Warden held several staff and command positions, including commander of Detachment 4 at Decimomannu, Italy (1984–85) and commander of the 36th Tactical Fighter Wing at Bitburg, Germany (1986–87). While in charge of the Warfighting Concepts Division at the Pentagon he developed the "Instant Thunder" plan, which became the foundation for the Operation Desert Storm air campaign and contributed to its successful execution. After the war, he served as the special assistant for policy studies and national security affairs to the vice president of the United States, and then as commandant of the U.S. Air Force Command and Staff College (1992–95). Warden received a BSc in national security affairs from the U.S. Air Force Academy in 1965 and an MA in political science from Texas Tech University in 1975. He was later selected to attend the National War College, where he wrote his well-known book *The Air Campaign: Planning for Combat* in 1985–86. He has published several articles, including "The Enemy as a System" (1995). Following retirement from active duty he became and remains the chairman and chief executive officer of Venturist, Inc., and is the coauthor of *Winning in Fast Time* (2002).

The Naval Institute Press is the book-publishing arm of the U.S. Naval Institute, a private, nonprofit, membership society for sea service professionals and others who share an interest in naval and maritime affairs. Established in 1873 at the U.S. Naval Academy in Annapolis, Maryland, where its offices remain today, the Naval Institute has members worldwide.

Members of the Naval Institute support the education programs of the society and receive the influential monthly magazine *Proceedings* or the colorful bimonthly magazine *Naval History* and discounts on fine nautical prints and on ship and aircraft photos. They also have access to the transcripts of the Institute's Oral History Program and get discounted admission to any of the Institute-sponsored seminars offered around the country.

The Naval Institute's book-publishing program, begun in 1898 with basic guides to naval practices, has broadened its scope to include books of more general interest. Now the Naval Institute Press publishes about seventy titles each year, ranging from how-to books on boating and navigation to battle histories, biographies, ship and aircraft guides, and novels. Institute members receive significant discounts on the Press' more than eight hundred books in print.

Full-time students are eligible for special half-price membership rates. Life memberships are also available.

For more information about Naval Institute Press books that are currently available, visit www.usni.org/press/books. To learn about joining the U.S. Naval Institute, please write to:

Member Services
U.S. Naval Institute
291 Wood Road
Annapolis, MD 21402-5034
Telephone: (800) 233-8764
Fax: (410) 571-1703
Web address: www.usni.org

